AF837953

# Understanding Blockchain in Construction

*Understanding Blockchain in Construction: A Practical Guide* is an essential resource that offers a comprehensive exploration of the transformative impact of blockchain technology in the construction industry. This book serves as a foundational reference, detailing how blockchain technology is the cornerstone for data-driven digital processes in the built environment, often referred to as the 'golden thread'. It aims to satisfy the growing need for knowledge and understanding among professionals in the enterprise, business, and information systems sectors of the construction industry, addressing both current capability needs and future opportunities.

This new addition to the Routledge Understanding Construction Series is a valuable resource for anyone in the construction industry looking to grasp how the data technology, blockchain, can revolutionise their work. In simple terms, it explains how blockchain can make construction projects more efficient, transparent, and sustainable. The book is filled with real-life examples showing how blockchain can help in various aspects of construction, like strict accountability to project safety rules and environmental standards, protecting architects' and engineers' original designs, and improving communication and data sharing between different software platforms used in construction. It's written in an easy-to-understand way, making it perfect for both beginners and experts. Essentially, this book is like a roadmap, guiding construction professionals through the exciting new world of blockchain, helping them stay updated and make their projects more effective and future-proof.

**Dr. Abel Maciel** is a practicing architect and Associate Professor at University College London (UCL). At UCL, he teaches courses on digital engineering management and digital transformation. His research interests encompass Digital Twins, Computational Design, Artificial Intelligence (AI), and Distributed Ledger Technology (Blockchain). Additionally, he serves as the Director of Design Computation Ltd, a specialised consultancy based in London, where his primary focus lies in the development and delivery of complex Building Information Modelling and Digital Twins, Programmatic Design, Digital Fabrication, and the exploration of new digital technologies for design, including Distributed Ledger Technologies.

## Understanding Construction

**Understanding NEC3: Engineering and Construction Short Contract**
*Kelvin Hughes* (2014)

**Understanding The Building Regulations 6th edition**
*Simon Polley* (2014)

**Understanding JCT Standard Building Contracts 10th edition**
*David Chappell* (2017)

**Understanding NEC4: Term Service Contract**
*Kelvin Hughes and Patrick Waterhouse* (2018)

**Understanding the NEC4 ECC Contract: A Practical Handbook**
*Kelvin Hughes* (2018)

**Understanding the NEC4 Professional Service Contract: A Practical Handbook**
*Kelvin Hughes* (2020)

**Understanding FIDIC: The Rainbow Suite**
*Kelvin Hughes* (2021)

**Understanding the NEC4 ECC Contract: A Practical Handbook, 3rd Edition**
*Kelvin Hughes* (2024)

**Understanding the NEC4 Term Service Contract: A Practical Handbook**
*Kelvin Hughes and Patrick Waterhouse* (2024)

**Understanding Blockchain in Construction: A Practical Guide**
*Abel Maciel* (2026)

# Understanding Blockchain in Construction

## A Practical Guide

Abel Maciel

Routledge
Taylor & Francis Group
LONDON AND NEW YORK

Designed cover image: © Shutterstock

First published 2026
by Routledge
4 Park Square, Milton Park, Abingdon, Oxon OX14 4RN

and by Routledge
605 Third Avenue, New York, NY 10158

*Routledge is an imprint of the Taylor & Francis Group, an informa business*



*British Library Cataloguing-in-Publication Data*
A catalogue record for this book is available from the British Library

ISBN: 978-1-032-85464-9 (hbk)
ISBN: 978-1-032-84784-9 (pbk)
ISBN: 978-1-003-51830-3 (ebk)

DOI: 10.1201/9781003518303

Typeset in Sabon
by Newgen Publishing UK

# Contents

# Figures

# Tables

# Preface

My involvement with blockchain in construction started at UCL Bartlett Faculty of the Build Environment and subsequently with the Centre for Blockchain Technologies (CBT). In the many meetings we have, we quickly realised there was a significant knowledge gap and potential in the construction industry, and we realised the need for a working group or consortium to bring together academia and the involved sectors to develop new technologies leveraging blockchain. The UCL Construction Blockchain Consortium was then established to develop a social network, research and support knowledge transfer, arrange commercial and academic presentations, assess and test commercial services and technology, conduct research, and drive policy, regulation, and understanding of the radical consequences of this technology and its associated services.

The purpose of this book is to provide an understanding of blockchain in the context of the construction industry. It aims to explain the technology's foundational philosophy, addressing both theoretical and practical aspects of blockchain applications, supplemented with relevant case studies.

This book is designed for a wide range of practitioners, including clients, contractors, subcontractors, and consultants. It is suitable for both those new to blockchain and those with more experience, offering valuable insights and practical guidance for all levels of expertise.

The content covers the evolution of digital transformation in the built environment, exploring how blockchain technology is reshaping this sector. It provides a comprehensive overview of the technological advancements and their impact on construction processes and outcomes.

The structure and layout of the book are organised to follow the technology from its simplest concepts to more complex elaborations. It discusses applications to construction projects and overarching concepts in the built environment, including practical examples and case studies to illustrate key points and enhance understanding.

This book is fully referenced and includes recommended additional materials and guidance. Please refer to the bibliography for further reading to deepen your understanding of blockchain in construction.

# Acknowledgements

I would like to express my deepest gratitude to my family. Their unwavering support, encouragement, and belief in me have been the cornerstone of my journey. Their love and guidance have provided me with the strength and determination to pursue my goals.

I am also immensely grateful to my research team at Bartlett Faculty of the Built Environment, University College London (UCL). Your dedication, insight, and collaborative spirit have been invaluable throughout this process. The countless hours of research, discussions, and problem-solving have significantly contributed to the depth and quality of this book. I am fortunate to have worked with such an extraordinary group of individuals.

Finally, I extend my heartfelt thanks to the editorial team. Your meticulous attention to detail, constructive feedback, and tireless efforts in refining this manuscript have been crucial. Your expertise and professionalism have greatly enhanced the clarity and coherence of this work, and for that, I am profoundly appreciative.

To all those who have supported me in this endeavour, both mentioned and unmentioned, your contributions have been instrumental in bringing this project to fruition.

# Introduction

## Blockchain Origins and Overview

The origins of blockchain technology can be traced back to early concepts of *decentralised systems*. These foundational ideas emerged from the fields of *cryptography* and *distributed computing*. Theoretical work in these areas laid the groundwork for decentralised ledgers by proposing methods for secure, decentralised information sharing without a central authority. Key developments in digital cash and cryptographic protocols in the 1980's contributed significantly to the evolution of decentralised systems (Chaum, 2024). These early ideas were crucial in shaping the principles that would later underpin blockchain technology.

As the need for secure communication grew, advancements in digital security became paramount. The evolution of cryptographic techniques, such as public-key cryptography and hash functions, played a critical role in enhancing data security and integrity. These techniques provided the tools necessary for creating secure communication protocols, which are essential for maintaining the confidentiality and authenticity of digital information. The development of these cryptographic methods was a key milestone that enabled the creation of secure, decentralised systems, paving the way for the development of blockchain technology.

Digital signatures emerged as a fundamental component in verifying authenticity and ensuring the integrity of digital transactions. By providing a means to validate the origin and content of a message or document, digital signatures laid the groundwork for many aspects of blockchain technology. They ensure that data has not been altered in transit and that the sender's identity can be confirmed. This capability is crucial for blockchain's operation, where trust and immutability are core principles. The introduction and widespread adoption of digital signatures significantly advanced the security and reliability of digital interactions.

Before the advent of Bitcoin, there were several notable experiments with digital currencies that sought to implement decentralised financial systems. Projects like DigiCash (Chaum, 1982) and e-Gold (Jackson, 1999) were early

DOI: 10.1201/9781003518303-1

attempts to create digital money that operated outside of traditional banking systems. While these early digital currencies faced various challenges and ultimately did not achieve widespread adoption, they provided valuable lessons. These included insights into the importance of security, user trust, and regulatory compliance. The successes and failures of these systems informed the design and development of subsequent blockchain-based cryptocurrencies, including Bitcoin, by highlighting the critical elements needed for a successful decentralised currency system.

## The Double-Spending Problem and Bitcoin

The double-spending problem is a fundamental challenge in digital assets and the transfer of units of value, like currencies and electronic transactions. It refers to the risk that a digital token or currency can be spent more than once, undermining its integrity and trustworthiness. Unlike physical money, digital information can be easily duplicated, making it possible for a malicious actor to fraudulently replicate and reuse the same digital asset multiple times.

Bitcoin, introduced by the pseudonymous Satoshi Nakamoto in 2008, marked the beginning of the cryptocurrency revolution because it solved the double-spending problem using blockchain technology. Nakamoto's seminal whitepaper, "Bitcoin: A Peer-to-Peer Electronic Cash System," outlined a vision for a decentralised digital currency that operates without the need for a central authority (Nakamoto, 2008). The key innovations introduced in this paper included the use of a decentralised ledger (blockchain) computing a consensus mechanism (proof-of-work in this case), and the elimination of double-spending, which collectively addressed critical issues in previous digital currency attempts.

Blockchain technology is the fundamental backbone of Bitcoin's functionality. It ensures the security and integrity of transactions through a decentralised and immutable ledger. Each transaction is recorded in a block, which is then linked to the previous block, creating a chain of blocks (blockchain), shown in **Figure 0.1**. This structure not only secures transaction data but also decentralises control, making the system resistant to censorship and fraud (Narayanan et al., 2016). The dual role of blockchain in Bitcoin is crucial: it ensures both the security of transactions and the decentralised nature of the network, allowing for thrustless interactions between parties.

## Cryptocurrency revolution

Following the introduction of Bitcoin, the cryptocurrency market experienced rapid growth and diversification. Bitcoin's success spurred the development of numerous alternative cryptocurrencies (altcoins), each aiming to improve upon or differentiate from Bitcoin in various ways. Key milestones

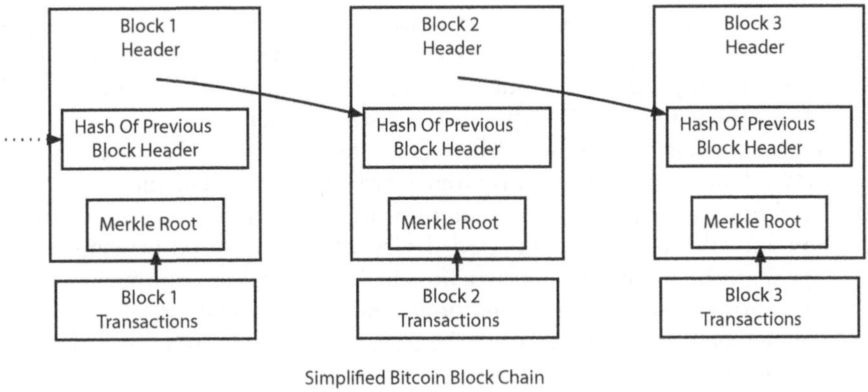


*Figure 0.1* **Simplified Bitcoin Blockchain.**

in the evolution of the cryptocurrency market include the launch of Ethereum in 2015, which introduced smart contracts (or a more sophisticated version of smart contracts in the blockchain), and the subsequent rise of decentralised finance (DeFi) applications (Buterin and Wood, 2013). This proliferation has led to a vibrant and expanding ecosystem, with cryptocurrencies gaining traction in mainstream finance and popular culture.

Despite its growth, the cryptocurrency market faces significant challenges and controversies. Regulatory hurdles and legal considerations are among the most pressing issues, as governments worldwide grapple with how to regulate and integrate cryptocurrencies within existing legal frameworks (Zohar, 2015). Additionally, scalability remains a major concern, with networks like Bitcoin and Ethereum experiencing congestion and high transaction fees during peak usage. Energy consumption is another critical issue, as the proof-of-work consensus mechanism used by Bitcoin requires substantial computational power, raising environmental concerns. Market volatility further complicates the adoption of cryptocurrencies as stable stores of value, posing risks to investors and users alike (Hong, 2016).

## Digital Construction and the effect of Blockchain

In a 'parallel universe', the construction industry has been evolving and taking small but decisive steps towards digital transformation. At this latest leap of industrial evolution in construction, digital transformation promises to improve efficiency, productivity, and sustainability. Key technologies at play include BIM, the IoT, Reality Capture & Reality Mesh, AI, and Blockchain. These innovations collectively contribute to modernising the construction industry by enabling more precise planning, real-time monitoring, predictive analytics, and secure data management (Oesterreich and Teuteberg, 2016).

**BIM** is a digital representation of the physical and functional characteristics of a building facility. It enhances project planning and execution by providing a comprehensive model that integrates various data sources. BIM allows for better collaboration among stakeholders, reduces errors, and improves project outcomes. When integrated with blockchain technology, BIM's data integrity and transparency are significantly enhanced.

**IoT** technology enables real-time monitoring and maintenance of construction sites through interconnected devices that collect and transmit data. Applications of IoT in construction include tracking equipment usage, monitoring environmental conditions, and ensuring worker safety. A critical challenge in IoT is securing the data exchanged between IoT devices, preventing unauthorised access and tampering.

**Reality Capture and Reality Mesh** technologies, such as 3D laser scanning and photogrammetry, create accurate digital models of physical environments. These technologies, combined with reality mesh, provide detailed and up-to-date representations of construction sites, facilitating better planning and management.

**AI and Machine Learning** are used in construction for predictive analytics, automation, and decision-making. These technologies help in optimising project schedules, predicting equipment failures, and automating routine tasks. AI applications in all sectors and with the advent of GenAI, there are numerous challenges, including ensures that the data fed into AI systems is reliable, which is essential for making accurate predictions and informed decisions (Shrouf et al., 2014).

Blockchain technology in the context of construction digital transformation promises to enhance transparency and trust in construction projects and business as usual by creating an immutable record of transactions. Each transaction recorded on the blockchain is permanent and cannot be altered, which ensures the integrity of the data. This capability is particularly beneficial for stakeholders who need to verify the authenticity and accuracy of project information. By providing a single source of truth, blockchain fosters greater trust and accountability among all parties involved in a construction project (Perera et al., 2020).

Therefore, blockchain can streamline supply chain management by tracking materials and components throughout the supply chain. It provides a transparent and traceable record of the origin, movement, and status of each item. This visibility helps in reducing fraud, delays, and inefficiencies by ensuring that all parties have access to the same accurate information. Additionally, blockchain can facilitate faster payments and contract enforcement, further enhancing supply chain efficiency (Kouhizadeh et al., 2021).

As blockchain becomes sophisticated and self-execute code and scripts, the technology can significantly improve contract management in the construction industry. By using blockchain, smart contracts automate and enforce agreements without the need for intermediaries. This automation reduces disputes and enhances compliance by ensuring that all contractual conditions are met before payments are released. Smart contracts also provide a transparent and tamper-proof record of all transactions, which helps in resolving disputes quickly and fairly (Mason and Escott, 2018).

This book provides an examination of business and information aspects in construction impacted by blockchain technology. It explores the fundamental principles of blockchain, offering a thorough understanding of how this technology works. The book addresses both the theoretical foundations and practical implementations of blockchain, ensuring that readers gain a holistic view of its potential and limitations in the construction sector. By covering these aspects, the book aims to equip readers with the necessary knowledge to understand and utilise blockchain technology effectively.

To illustrate the impact of blockchain, this book presents a series of real-world examples and case studies. These case studies showcase blockchain projects failures and successes, highlighting practical insights and lessons learned. It also examines current trends and challenges, identifying new directions and innovations that are shaping the future of the industry. This approach not only demonstrates the potential benefits but also addresses the challenges encountered and strategies employed to overcome them.

On the development of the text, I've tried to make it accessible to the wide range of practitioners, including clients, contractors and technology development functions. It aims to provide value to both newcomers and experienced professionals. For those new to blockchain. The book's relevance to a diverse audience ensures that it serves as a valuable resource for anyone involved in the construction industry.

The book is structured to guide readers logically from basic concepts to advanced applications. It begins with an introduction to the fundamental principles of blockchain, followed by detailed discussions of its practical applications related to construction. The inclusion of short summaries in the beginning of each chapter, practical examples, case studies and refences for further reading allow for further learning on each topic. This structured approach ensures that readers can build their understanding progressively, gaining confidence as they move through the book.

Finally, by promoting this nascent technology and new paradigm, the book aims to inspire industry professionals to embrace blockchain technology and integrate it into their practices, equipping readers with the knowledge and tools to navigate and leverage this disrupting foundation for new models of practice in construction.

Chapter 1

# Foundational Concepts of Blockchains

## 1.1 Network Systems

Networks are configurations of interconnected entities, known as nodes, linked by relationships or edges. These structures are pervasive, encompassing systems such as social interactions, biological pathways, and technological infrastructures, including DLT and therefore, blockchains. The study of these intricate interconnections is termed network science, an interdisciplinary field that integrates principles from mathematics, physics, computer science, as well as sociology to analyse and interpret network behaviour.

Networks underpin the functionality of numerous systems in our daily lives, influencing how information disseminates, how diseases propagate, and the robustness of various infrastructures to disruptions. Insights from network science facilitate the optimisation and resilience of these systems, thereby enhancing adeptness and stability across diverse domains.

To effectively understand blockchain systems, it is essential to examine different network structures and their characteristics. This section discusses four primary network architectures –centralised, decentralised, distributed, and hybrid – starting with the simplest form: the centralised network.

### *1.1.1 Centralised Networks*

In a centralised network, all data and control are managed by a single central server or authority. Peripheral nodes (clients) must communicate through this central point to access resources or services (Figure 1.1). This structure simplifies management but introduces a single point of failure, making the network vulnerable if the central server encounters issues.

### *1.1.2 Decentralised Networks*

Decentralised networks distribute control across multiple nodes, eliminating reliance on a single central authority (Figure 1.2). Each node operates independently yet collectively contributes to the network's functionality. This

DOI: 10.1201/9781003518303-2

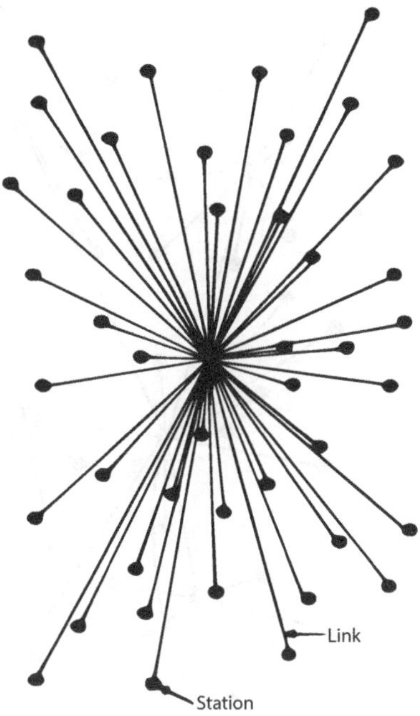


*Figure 1.1* Centralised Network Model.

arrangement enhances fault tolerance and reduces the risk of total network failure, as the system can continue to function even if some nodes fail.

### *1.1.3 Distributed Networks*

Distributed networks consist of interconnected nodes that share data and computational tasks, working collaboratively to achieve common objectives (Figure 1.3). Unlike in centralised systems, no single node holds overarching control in distributed networks. This configuration enhances scalability, fault tolerance, and resource utilisation, as tasks are distributed across multiple nodes, reducing the impact of individual node failures.

Understanding these architectures is crucial in the understanding of blockchain, where implementing decentralised or distributed networks can improve coordination of information by enhancing data accessibility, collaboration, and resilience against system failures.

### *1.1.4 Hybrid Networks*

More recently, hybrid network architectures exist and are employed to leverage the advantages of multiple networking models combined in one

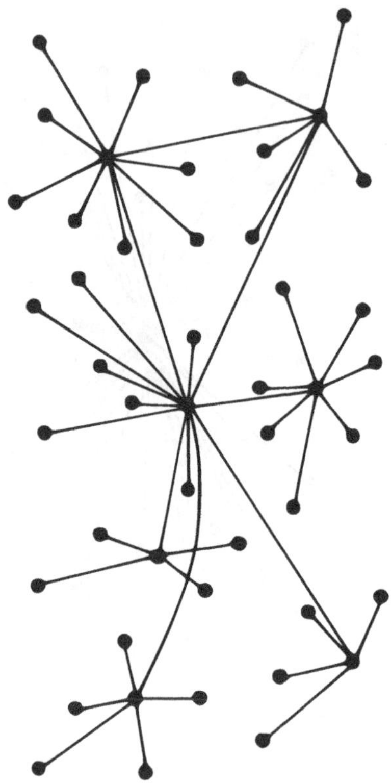

*Figure 1.2* Decentralised Network Model.

system. By integrating elements and topologies from centralised, decentralised, and distributed networks, hybrid networks are both robust and efficient. For example, a hybrid network might integrate on-premises data centres with public cloud services, effectively merging traditional centralised systems with cloud-based distributed architectures. This approach allows organisations to create optimised network architectures to meet their capability needs in respect to performance, security, scalability and resilience by utilising the strengths of each integrated network type.

## 1.2 Distributed Ledgers Technologies

DLT refers to decentralised data management systems where multiple network nodes maintain and synchronise a shared ledger of transactions, removing the need for a central administrator. Key features of DLT typically include decentralisation (no single point of control), immutability (entries are tamper-resistant), and transparency (participants can verify ledger contents). However, not all DLT implementations necessarily include all these attributes.

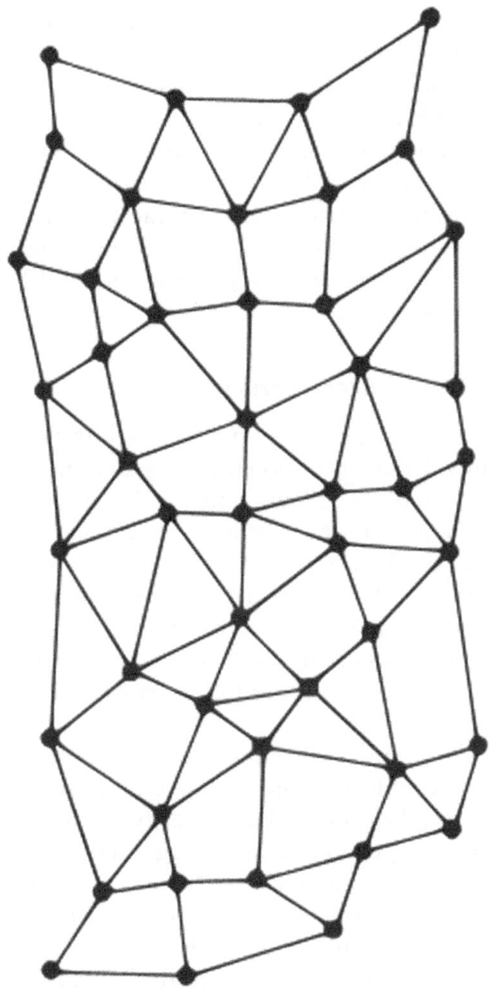

*Figure 1.3* Distributed Network Model.

When properly architected, DLT enhances trust and security by employing cryptographic techniques that secure records, making fraud and unauthorised alterations highly difficult. Although DLT handles ledger data, it is generally unsuitable as a traditional database, since synchronising large volumes of data would be inefficient. Instead, DLT systems often store references to external data, providing greater resilience—no single point of failure—and ensuring improved data integrity. Altering records requires compromising many nodes simultaneously, making attacks computationally prohibitive.

This decentralised approach offers significant advantages over centralised ledgers, such as distributing risk to minimise impact from node failures or malicious activity. Cryptographic security ensures past transactions are nearly impossible to alter without detection, and transparency facilitates auditing and accountability, particularly useful in multi-party processes like finance and supply chains. DLT also enables peer-to-peer transactions and automated agreements through smart contracts, reducing the need for intermediaries, increasing processing speed, and lowering costs. Moreover, as participation only requires an internet connection, DLT has the potential to enhance financial inclusion.

### *1.2.1 DLT Network Architecture*

DLT networks employ a peer-to-peer (P2P) architecture where all nodes are equal participants. Unlike traditional client-server models – where a central server responds to multiple clients – in a P2P network each node acts simultaneously as client and server, directly communicating with other nodes. With no central coordinator, every node can independently initiate transactions and distribute data. This decentralised mesh topology significantly enhances fault tolerance, enabling the network to remain operational even if many nodes fail, provided that sufficient peer connections remain intact.

Nodes within a P2P network can fulfil various roles:

- **Full Nodes:** These nodes store the entire ledger, including all blocks and transactions, and independently validate new entries. Acting as the network's backbone, they ensure data integrity by checking digital signatures, preventing double-spending, and propagating validated updates to other nodes. Public blockchains rely heavily on maintaining enough full nodes to ensure decentralisation and resistance to censorship.
- **Light Nodes:** Also known as lightweight or Simplified Payment Verification (SPV) nodes, they do not store the entire ledger. Instead, they retain only critical information—such as block headers—necessary to verify their own transactions, relying on full nodes to supply detailed data.
- **Miner/Validator Nodes:** These specialised nodes store the full ledger and perform computationally intensive tasks required to solve cryptographic puzzles and validate blocks within the DLT.
- **Other Node Types:** Certain DLT networks specify additional node roles. For instance:
  - **Masternodes** (common in some cryptocurrencies) are full nodes providing extra functionalities, such as coordinating mixing services or governance decisions, often requiring collateral deposits.
  - **Pruned Nodes** store only recent ledger history, removing older data to conserve storage while still validating new transactions.
  - **Archive Nodes** maintain the complete historical ledger, enabling queries into distant past transactions.

Nodes within a DLT network typically follow a shared protocol for peer discovery, transaction broadcasting, and data validation. For example, Bitcoin employs a randomly wired network where nodes establish arbitrary connections without a fixed topology. By default, a Bitcoin node initiates eight outgoing connections to randomly selected peers while also accepting incoming connections, creating a robust ad-hoc mesh. This structure ensures redundancy by propagating data through multiple independent pathways, eliminating any single point of failure.

### *DLT Communication Protocols*

In a DLT network, nodes achieve data consistency through constant communication. When a new transaction is created (e.g., Alice transfers ownership of coins to Bob), it's broadcast to the network so that all nodes eventually receive and record it. This dissemination usually happens via a **gossip protocol**, which works similarly to how rumours spread in a crowd: each node randomly forwards the information to some of its peers, who in turn forward it to others, and so on. Over a short time, the transaction disseminates its way to every corner of the network. This epidemic-style propagation is highly robust – even if some nodes fail to forward it, others will ensure the message gets through. It does, however, introduce redundancy (many nodes may receive the same transaction from multiple peers), which is a trade-off for reliability.

### *DLT Consensus Mechanisms*

Consensus mechanisms are the protocols by which distributed nodes agree on the contents of the ledger. In a network with no central authority, consensus is vital to ensure that all honest nodes converge on the same one true history of transactions, even if some nodes fail or act maliciously. Different blockchains use different consensus algorithms, each with trade-offs in security, efficiency, and network overhead.

It is important to note that DLT refers broadly to distributed systems where data is replicated across multiple network nodes and synchronised through consensus algorithms. DLT acts as a shared system of records among participants, which can be permissioned (controlled access) or permissionless (open access). A blockchain is a specific type of DLT that organises data into a continuous chain of blocks using cryptographic techniques. While every blockchain is a form of DLT, not every DLT employs a blockchain architecture. Distributed ledgers can use various other structures and consensus mechanisms beyond this linear, hash-linked approach.

## 1.3 Blockchain Architectures

Blockchain technology organises transactions into groups known as blocks, which are sequentially appended to form a linear chain. Each block contains transaction data and associated metadata, including a cryptographic hash pointer that securely links it to the preceding block.

If the transaction data in an earlier block is modified, its hash changes accordingly. Since each block stores the hash of its predecessor, altering data in one block invalidates the hashes of all subsequent blocks. This property immediately reveals any attempt at tampering, thereby ensuring the integrity and immutability of the ledger and creating a robust, secure, chronological record of transactions.

Unlike conventional databases – which are typically managed by a single authority (such as a cloud provider) – blockchain technology is truly distributed, meaning no single participant controls the ledger. Therefore, blockchains are not direct replacements for traditional databases; rather, they offer a method for securely managing transactions or records among multiple independent participants. As such, describing blockchains merely as "distributed databases" can often be misleading.

### *1.3.1 Network Architecture*

Blockchain implementations operate using a peer-to-peer (P2P) network architecture, where many nodes retain a full copy of the ledger and participate actively in transaction validation and consensus processes. For example, in public blockchains such as Bitcoin or Ethereum, every node independently validates each transaction and block. Communication typically occurs through unstructured peer-to-peer connections (nodes connect randomly with multiple peers) and employs a 'flooding' or 'gossip' protocol, meaning whenever a new block is created ("mined"), it is broadcast across the entire network rapidly.

This network structure provides several critical benefits:

- **Redundancy**: Every node independently validates transactions, reducing risks of centralised failure or manipulation.
- **Transparency:** The ledger contents are fully visible to all participants, allowing easy auditing and verification.
- **Consistency:** Nodes generally have homogeneous roles, meaning each agent in the network equally contributes to ledger validation and synchronisation. Exceptions exist in specialised blockchain systems where certain nodes handle distinct roles, such as miners in Bitcoin or validators in proof-of-stake systems.

However, this uniform, distributed design comes with trade-offs, including potential inefficiencies, particularly for scenarios that require high throughput or large volumes of data synchronisation. Permissioned blockchains mitigate some inefficiencies by restricting network participation to selected nodes, but

```
Block #101:

{
"block_number": 101,
"timestamp": "2024-05-10T12:34:56Z",
"transactions": [
"Alice sends 3 BTC to Bob",
"Charlie sends 1 BTC to Dave"
],
"previous_block_hash":
"0000000000000000000234abef1234567890abcdef1234567890abcdef123456",
"current_block_hash":
"00000000000000000000789abcd4567890123abcdef4567890123abcdef789012"
}
```

*Figure 1.4* Block Hash Sample.

typically still follow similar network architectural principles, maintaining redundancy and ensuring consistent ledger updates.

### *1.3.2 Data Architecture*

We have previously seen that in a blockchain, transactions are grouped into blocks, which are then appended sequentially to form a single, continuous chain. Each block contains:

- **Transaction data:** A batch (list) of recent transactions.
- **Metadata:** Includes a timestamp (indicating when the block was created) and a cryptographic hash pointer to the previous block.

This forms a linear, sequential structure – the "blockchain" – in which each block directly extends from the preceding block. Because of the cryptographic linkage, altering any data in an earlier block changes its hash, causing mismatches in subsequent hashes. Such a mismatch immediately highlights any tampering, preserving the integrity of the ledger. This sequential design ensures a robust, tamper-resistant historical record of all transactions but inherently processes updates sequentially, one block at a time.

In the above example, Figure 1.4, the *previous_block_hash* connects block #101 to block #100, ensuring chronological order and immutability.

#### *Using Cryptographic Hashing*

In a traditional database, data can be modified or deleted by those with access, which can lead to potential security risks or errors. In contrast, the blockchain cryptographic hashing creates a unique and permanent record for each transaction. Once a transaction is added to the blockchain, it cannot be altered or erased, providing a high level of integrity.

```
import hashlib

# The input text
input_text = "A"

# Calculate the SHA-256 hash
hash_result = hashlib.sha256(input_text.encode()).hexdigest()

# Print the resulting hash
print(f"SHA-256 hash of '{input_text}': {hash_result}")
```

*Figure 1.5* SHA-256 Sample.

A hash function takes any input and transforms it into a unique, fixed-length output called a **hash**, which appears random. Crucially, even a minor alteration to the input drastically changes the resulting hash, making any changes immediately evident. For example, a hash created using **SHA-256** (Secure Hash Algorithm 256-bit) would look like this:

- Input: the words "Construction Blockchain"
- Hash output: 27a2b929c7ee839764a2bcf728deb5dcdff8b74e1719d71df-10842c9a97c7e62
- Input: the letter "A"
- Hash output: ca978112ca1bbdcafac231b39a23dc4da786eff8147c4e72b9807785afee48bb

The source code to generate the **SHA-256** hash of the letter "A" using Python programming language and libraries would look like Figure 1.5:

Thus, cryptographic hashing ensures the integrity of the blockchain, as each block's hash depends directly on its data and the previous block's hash. Any modification breaks the consistency of the entire chain.

### *1.3.3 Consensus Mechanism*

All DLTs, including blockchains, rely on consensus mechanisms to validate and order transactions. Consensus algorithms ensure all nodes agree on the ledger state, preventing conflicting transactions.

#### *Proof-of-Work and Proof of Stake*

Public blockchains often rely on Nakamoto-style consensus which uses economic incentives and competition. The classic example is Proof-of-Work (PoW), where miners (running mines' nodes in the network) compete to solve a cryptographic puzzle; the winner with the fastest hardware kit creates the next block and is rewarded. PoW is robust but resource-intensive, requiring significant computation to add each block. Newer blockchains use Proof of Stake (PoS), where validators stake cryptocurrency and are randomly chosen (weighted by stake) to produce blocks. PoS avoids heavy computation but still generally progresses one block at a time.

Both PoW and PoS are probabilistic consensus mechanisms – 'forks' in the network can occur, but eventually one chain becomes the accepted ledger (with finality achieved after multiple confirmations). There are also hybrids and variants (e.g. Delegated PoS, Proof of Authority for private chains), but they all revolve around agreeing on the next block in the chain. In summary, blockchain consensus tends to be block-based: reaching agreement on each new block sequentially, often incurring some delay to ensure the block is widely propagated and agreed upon.

#### *Byzantine Fault Tolerance*

Many non-blockchain DLTs use more traditional distributed consensus algorithms, often inspired by Byzantine Fault Tolerance (BFT) research. For example, permissioned DLT networks might use PBFT (Practical Byzantine Fault Tolerance) or its variants, where nodes exchange votes in multiple rounds to agree on each transaction or batch. These are voting-based consensus mechanisms – they achieve final agreement through a fixed number of communication rounds among a known set of validators. Such BFT protocols can finalise transactions faster (in seconds) and with absolute finality (once agreed, it's final), but typically require a limited number of trusted validators to be efficient.

#### *Raft, Paxos and Variations*

Another approach in some DLTs is leader-based consensus using protocols such as Raft or Paxos (common in distributed databases). In these protocols, a designated leader (which can rotate or be elected) orders transactions and followers to replicate them; this provides crash fault tolerance (CFT) in closed networks. For instance, Linux Foundation's Hyperledger Fabric open-source technology ordering service can use Raft (a CFT protocol) to agree on transaction order within a permissioned network. In a few ledgers using Directed Acyclic Graphs (DAG), consensus on transaction uniqueness is achieved via a cluster of notary nodes, which may run a BFT algorithm; only the notaries (not all participants) run the consensus to confirm that no double-spend occurred. DAG-based ledgers also have unique consensus styles, sometimes using a probabilistic algorithm where 'tips' (unconfirmed transactions) are randomly selected for confirmation – consensus emerges from the cumulative weight of validations (and previously involved a coordinator for finality). 'Hashgraph' uses an asynchronous BFT algorithm: nodes gossip transactions and then use virtual voting to deterministically agree on an order of events, achieving consensus very fast without mining.

*Table 1.1* Consensus Algorithms Categories

| *Mechanism* | *Description* |
|---|---|
| PoW/PoS (Blockchain) | Decentralised competition (mining/forging) to extend the chain – e.g. Bitcoin's PoW, Ethereum's PoS. |
| PBFT and Variants (DLT/BFT) | Rotating leaders and multiple voting rounds among validators – e.g. Hyperledger Fabric's consensus or certain consortium chains. |
| Leader-based (DLT/CFT) | A single leader ordering transactions at a time – e.g. Raft in some private chains. |
| Specialised (DAG-based) | Hashgraph's gossip and virtual voting, IOTA's tip selection – designed for high throughput without mining. |

Thus, consensus algorithms in DLT are chosen based on network model: open blockchains vs. consortium or private ledgers use different techniques to achieve trust. Table 1.1 summarises typical examples:

### *1.3.4 Immutability and Security*

A critical promise of blockchain and DLTs is **data integrity** – once a transaction is accepted, it cannot be tampered with or erased without detection. Different DLT architectures achieve this immutability through distinct mechanisms. In blockchains, a cryptographic chain of blocks, reinforced by distributed consensus, makes the ledger both tamper-evident and tamper-resistant.

In public blockchains, security is further strengthened by economic incentives. Altering historical data would require redoing the PoW for all subsequent blocks or controlling most of the staking power or hashpower—both of which are prohibitively difficult and nearly impossible. This architecture has proven highly robust; for example, Bitcoin's ledger has never been maliciously altered, as the collective honest hash power has consistently outpaced attackers.

Additionally, blockchains use a binary hash tree or **Merkle trees** within each block, where all transactions are hashed together. A Merkle tree (Figure 1.6) is a data structure that organises transactions into a tree of hashes, enabling efficient and secure verification of data. Even a minor change to a single transaction alters the Merkle root, which in turn changes the block's hash, making tampering immediately apparent. The distributed nature of the network – where many nodes independently store the full ledger – ensures that any falsified copy is rejected by the majority. As a result, blockchains achieve immutable, append-only logs: new blocks can be added, but existing records cannot be altered.

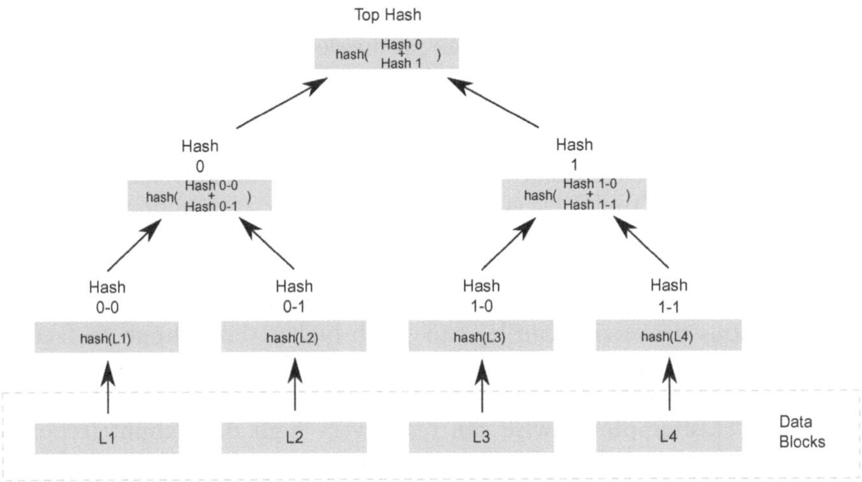


*Figure 1.6* An example of a Merkle tree.

In blockchain systems, security derives primarily from this decentralisation and cryptography. As long as no single entity controls most consensus power and cryptographic algorithms remain sound, the history remains secure. In contrast, other DLTs using BFT rely on a different trust model: security depends on a quorum of pre-selected, trusted nodes acting honestly and not colluding.

### 1.3.5 *Scalability and Performance*

The architecture of a ledger significantly impacts its throughput (transactions per second), latency, and scalability (ability to grow with more users or load).

The linear, sequential nature of blockchains means that transactions are essentially processed in batches (blocks) one after the other. Even if multiple miners or validators are working in parallel, ultimately only one block gets added to the chain at a time (per interval). This creates a bottleneck, because no matter how many nodes you add, the blockchain's throughput is limited by block size and block time (or gas limits in systems like Ethereum). For example, Bitcoin averages seven transactions per second and Ethereum 10-30 TPS in their base layer, because of these protocol limits. Increasing throughput often requires changing parameters (bigger blocks, faster blocks), which can hurt decentralisation or security (e.g., propagating large blocks slower, giving an edge to certain miners). Thus, traditional blockchains have struggled with scalability – as usage grows, networks have seen congestion, high fees, and slower confirmations. Efforts like the Lightning Network (layer-2) or Ethereum's sharding upgrade aim to overcome these limits, but at the base layer, the sequential block production is a rate-limiter.

Other DLT designs attempt to improve performance by allowing parallel processing or reducing consensus overhead and they may one day be adopted by blockchain. For example, DAG Ledgers don't require a single global block sequence, they can, in theory, confirm transactions in parallel. In IOTA's Tangle, every node that issues a transaction also helps validate others, so many transactions can be validated simultaneously across the network. There is no block size constraint – each transaction is its own record – and no fixed interval to wait for a new block. This asynchronous processing can lead to much higher throughput. In fact, as more users transact (and thus participate in validation), the capacity of the Tangle can increase. DAG-based networks often tout very high TPS potential. Hashgraph likewise can reach very high throughput (reported hundreds of thousands of TPS in controlled settings) because it leverages fast gossip and does not slow down for block intervals. Consensus finality in Hashgraph is achieved quickly (often within a few seconds) since virtual voting computes the results without multiple on-chain rounds. Nano, using a BlockDAG architecture, allows each account to have its own blockchain (account-chain), enabling transactions to be confirmed individually and quickly, achieving high TPS.

In a controlled DLT environment (like a consortium), consensus algorithms like PBFT can finalise a batch of transactions in one round of messages. These systems can reach hundreds or thousands of TPS on normal enterprise hardware because they don't rely on PoW delays or global broadcast to all nodes. For example, Hyperledger Fabric has achieved over 3,000 TPS in tests by limiting endorsement and ordering to a subset of nodes. Corda's performance can scale with the number of bilateral interactions – since not all nodes see all transactions, adding more nodes doesn't burden each node with all traffic. The scalability bottleneck in Corda is often the notary service; however, notaries can be clustered or sharded by different asset types to distribute the load. Essentially, private DLTs can scale horizontally by partitioning the workload (different sets of nodes handle different transactions), something a global blockchain cannot easily do without sharding.

The sequential block production in blockchains also impacts latency – a transaction might have to wait for the next block, which could be minutes (Bitcoin ~10 min per block) or seconds (Ethereum ~12 seconds). By contrast, DAG or BFT systems often have faster finality – a transaction might be confirmed as soon as enough votes are collected or references accumulated (potentially in real-time). For instance, Hedera Hashgraph's architecture allows it to finalise transactions in a few seconds with high confidence, whereas a similar confirmation on a PoW blockchain might require waiting multiple block intervals (minutes) for finality. This makes DAG and other designs attractive for high-frequency or Internet of Things (IoT) use cases where quick, scalable processing is needed.

However, trade-offs exist. High-throughput DAG networks might face challenges in maintaining security/decentralisation when scaling (e.g., the need for a coordinator in IOTA's early days was partly to protect the ledger when activity was low). Similarly, BFT consensus doesn't scale to hundreds of validators easily – it works best with tens of nodes, which is fine for permissioned settings but not applicable to open networks without sacrificing decentralisation. Blockchains sacrifice some performance for a more open participation model. Recent innovations also blur lines: Ethereum's move to PoS and sharding, Layer 2 networks, and sidechains aim to improve blockchain scalability, while some DAG projects introduce fixed checkpoints or "blocks" to improve stability. So, each approach addresses scalability differently, but generally, blockchain = simpler, slower; other DLTs = complex, faster. As one source notes, DAG ledgers are often seen as more scalable due to parallel validation and lack of mining, often achieving higher throughput with lower fees than traditional blockchains.

## 1.4 Tokenisation and Smart Contracts

Blockchain technology enables solutions like tokenisation and smart contracts to creating secure, automated, and decentralised systems for asset management and transactions.

Tokenisation is the process of converting real-world assets into digital tokens on a blockchain. In this process, ownership rights to an asset (whether physical like real estate or intangible like equity) are recorded as tokens on a distributed ledger. Each token typically represents a share or stake in the underlying asset, enabling fractional ownership – multiple investors can own a portion of an asset that previously might have been too costly or illiquid to split. By breaking assets into smaller digital units, tokenisation increases liquidity: these tokens can be traded more easily and frequently on digital marketplaces, unlocking value from assets (e.g. property or art) that traditionally take a long time to sell. It also brings transparency and security: transactions and ownership are tracked on an immutable blockchain ledger, which all parties can verify, reducing fraud and the need for intermediaries. In short, tokenisation democratises access to investments by lowering barriers to participation and allowing 24/7 peer-to-peer trading of asset shares globally. Businesses benefit through access to a wider investor pool, more efficient capital raising, and the ability to unlock value from illiquid holdings.

Smart contracts are self-executing digital agreements stored on a blockchain, which automatically run when predetermined conditions are met. Think of them as computer programs that enforce contract terms: "*if X condition happens, then perform Y action.*" For example, a smart contract could release a payment automatically when a shipment is delivered, without requiring escrow or paperwork. These contracts run on decentralised networks, so once deployed they operate **without intermediaries** – participants can be confident the terms will execute fairly without needing a bank, lawyer, or other middleman to intervene. This **trust model** relies on blockchain transparency and cryptography: all parties can see the code and outcomes, and no single entity can alter the record, providing assurance that the agreement can't be tampered with or unfairly changed. Smart contracts offer **security advantages** as well. The code and transaction records are encrypted and distributed across the network, making them resistant to hacking or fraud (altering one record would require altering the majority of copies across the network). They also enhance efficiency: execution is **automatic and instant** once conditions are satisfied, which reduces delays and human error. In summary, smart contracts enable reliable, automated agreements in a trust-minimised way, which can streamline business processes and reduce costs by replacing manual oversight with code.

### *1.4.1 How Tokenisation and Smart Contracts Work Together*

Most tokenisation today happens on smart contract platforms like Ethereum. Ethereum introduced standard protocols (token standards) that ensure tokens are created and managed in a consistent way. For example, ERC-20 is the technical standard for fungible tokens on Ethereum. Fungible tokens are interchangeable – each unit is identical in value and properties (like one share of a stock, or one USDC coin is the same as any other). ERC-20 defines a common set of functions (like transferring tokens or checking balances) so that any wallet or exchange can handle any ERC-20 token reliably. This standard underpins most ICO tokens, stablecoins, and other cryptocurrencies. By contrast, ERC-721 is a standard for non-fungible tokens (NFTs) which represent unique assets. An ERC-721 token has an identifier that makes it distinct from all others; this standard defines how to create, transfer, and track ownership of one-of-a-kind tokens (for example, a specific digital art piece or collectable). There's also ERC-1155, a newer multi-token standard that can handle both fungible and non-fungible items in one smart contract. ERC-1155 is useful in scenarios like gaming platforms, where a single contract might manage many types of tokens (e.g. gold coins that are fungible and unique sword items that are non-fungible. These standards are important for businesses because they ensure interoperability – a token issued following these conventions can be supported by existing blockchain infrastructure (wallets, trading platforms, etc.) without custom integration. Beyond Ethereum, other blockchains have analogous standards (for instance, Binance Smart Chain's BEP-20 is similar to ERC-20). When implementing tokenisation, choosing the right token standard depends on whether the asset is meant to be divisible and uniform (use a fungible token standard) or unique (use an NFT standard).

Industries that adopt blockchain early will gain a competitive advantage, while those that ignore it risk being left behind in a digitally transformed economy. Whether it's finance, real estate, supply chains, healthcare, or digital commerce, tokenisation and smart contracts are creating new efficiencies, investment opportunities, and business models. Forward-thinking companies should explore blockchain to future-proof their operations and tap into the next wave of innovation.

## 1.5 Notable Blockchain Case Studies

To illustrate the architectural differences, consider a few prominent Blockchains and DLT implementations **beyond blockchain**, and how their design diverges:

- **IOTA's Tangle (DAG):** IOTA is a distributed-ledger designed for the IoT, using a DAG called the **Tangle** instead of a blockchain. In the Tangle, there are *no miners and no blocks*. Every new transaction must approve two existing transactions by referencing their hashes. This creates a tangled graph of transactions. Consensus is inherently achieved by each transaction validating two previous ones; a *Coordinator* node was used (until recently) to finalise milestones for security, but the goal is a completely decentralised DAG. The absence of blocks means IOTA has **no fixed transaction fees** (since adding a transaction only requires doing a small amount of PoW to help approve two others) and theoretically unlimited scalability – as more devices use it, they contribute validation power. Architecturally, this is very different from a blockchain: there's no global chain or ordering enforced by miners; instead, different parts of the DAG may grow simultaneously and eventually merge when transactions reference one another. The Tangle is suitable for high-throughput, microtransactions (e.g. sensor data payments) because of its parallel nature and feeless design. The trade-offs include more complex tip-selection algorithms and the need for mechanisms to prevent conflicting subtangles. *Difference from blockchain:* Tangle's DAG structure allows **parallel transaction confirmations** and removes miners, whereas blockchains serialise confirmations block by block.
- **Hedera Hashgraph:** Hashgraph is a patented DLT technology (underlying Hedera network) that uses a gossip protocol and **virtual voting** to achieve Byzantine consensus extremely fast. Instead of a chain of blocks, Hashgraph builds a graph of gossip events. When nodes gossip, they include hashes of the last events they know ("gossip about gossip"), which allows nodes to eventually know the entire history of gossip exchanges. Using this information, every node can independently calculate the consensus order of transactions via virtual voting without actually sending votes (hence saving communication). The result is **asynchronous BFT**

consensus with high throughput (Hedera Hashgraph touts thousands of TPS) and rapid finality (in seconds). Hashgraph's data structure is essentially a **DAG of events** – each event contains transactions plus references (hashes) to earlier events. This differs from a blockchain where new data is only appended at one end of a single chain. In Hashgraph, many events can be added all over the place concurrently, and the algorithm still yields a single agreed history. Hedera's network is permissioned-public (a known set of governors running nodes, but open for anyone to use). *Difference from blockchain:* Hashgraph lacks a linear chain – **there are no blocks.** Security doesn't come from PoW or longest chain, but from math (BFT voting). It achieves similar goals (decentralised agreement on transactions) but via a radically different technique, favouring speed and fairness (each transaction gets a consensus timestamp when 2/3 of nodes have received it, and ordering is by that timestamp) over the simpler but slower block-by-block approach

**R3 Corda:** Corda is often described as a "DLT platform" rather than a traditional blockchain. It was built for financial institutions to record contracts and transactions in a shared system **without a global blockchain**. In Corda, data is shared on a **need-to-know basis** – only parties to a transaction (and relevant regulators or oracles, if permitted) receive the transaction details, rather than broadcasting to the whole network. Consensus in Corda is twofold: *validity* (parties agree transaction is correct, signatures verify, etc.) and *uniqueness* (a notary service assures no double-spend of the same asset or contract state). Multiple notary clusters may exist for different asset types or business networks. The notaries are the only semi-centralised component: they act as authorities to confirm a transaction's uniqueness. They can run BFT algorithms among themselves if decentralised trust is needed (so that no single notary node is a point of failure). The data structure in Corda is UTXO-like: transactions consume previous state objects and produce new state objects, with cryptographic hashes linking them. This forms a graph of transactions for each piece of data, but no single chain ties all transactions together. Immutability is per state/contract – once final, a transaction's outputs can only be changed by another valid transaction consuming them (with sign-off by required parties). *Difference from blockchain:* Corda sacrifices the global visibility and total ordering of a blockchain in favour of privacy and scalability in a permissioned environment. There's no native token or mining; consensus is achieved by business logic and trusted notary services. It's effectively a distributed database of facts, with parties only sharing the facts they need, secured by signatures and hash-links. This makes Corda suitable for industries like banking where privacy is paramount and participants are known, but it's not suitable for a trustless public cryptocurrency (by design, it's not aiming for permissionless use).

Others: There are several other non-blockchain DLT projects. *Examples:* Holochain takes an agent-centric approach, where each participant has their own chain and there is no global ledger – data is shared via a DHT (distributed hash table) rather than all nodes agreeing on one sequence. This drastically different architecture aims at massively scalable and user-controlled data, but it's a departure from the typical ledger model. Hashgraph (Swirlds) we covered via Hedera. Nano (formerly RaiBlocks) uses a Block-Lattice: each account has its own blockchain, and consensus is achieved through voting on conflicting transactions – this is another DAG/graph structure enabling fast, feeless transfers for a cryptocurrency. Hyperledger Fabric (while sometimes called a blockchain) actually has a modular consensus and doesn't strictly build an unbroken chain visible to all – it uses channels and an order-execute architecture that's more akin to a distributed database with cryptographic audit trail. These examples highlight that DLT comes in many flavours, each tailored to certain use cases: Tangle for IoT microtransactions, Hashgraph for high-speed corporate or public networks, Corda for financial agreements, etc.

## 1.6 Evolution of Blockchain Maturity

Blockchain technology has matured dramatically since Bitcoin's debut in 2008, evolving through distinct layers of capability. What began as a peer-to-peer digital currency network has grown into platforms for smart contracts, scalable off-chain solutions, enterprise applications, and decentralised finance. This session outlines key stages of blockchain development from 2008 to 2025, relates them to Gartner's Hype Cycle, and discusses major technologies built on blockchain (e.g. DeFi, NFTs, DAOs, Web3, tokenisation, digital identity, supply chain applications).

Blockchain's trajectory maps closely to Gartner's Hype Cycle. By 2016–17 it had reached the Peak of Inflated Expectations, amid widespread hype that it would revolutionise industries (Perez, 2019). The crash of 2018 then led to a Trough of Disillusionment: in 2019 Gartner noted the technology had "not yet lived up to the hype" and many initiatives failed. Since then, it has slowly climbed the Slope of Enlightenment with improvements in scalability and real pilots. Gartner predicted that by 2023 blockchain platforms would be much more scalable and interoperable (Perez, 2019). As of 2025, some blockchain applications are finally approaching a Plateau of Productivity while others remain in development.

### *1.6.1 Key Innovations Built on Blockchain*

- **Decentralised Finance (DeFi):** Financial services run through smart contracts on public blockchains. Users can lend, borrow, and trade crypto assets without banks. By 2022, DeFi platforms held $200 billion in assets (Bender, 2022).

- **NFTs:** Unique tokens representing ownership of digital assets (artwork, video, virtual item, etc.). NFT trading surged in 2021 to about $25 billion (Howcroft, 2022), proving that blockchain can assign verifiable ownership and provenance to digital goods.
- **Decentralised Autonomous Organisations (DAOs):** Communities or organisations governed by blockchain-based rules and token-holder votes instead of centralised management. DAO membership exploded (from ~13k to 1.7 million people in 2021) (WEF, 2022), enabling internet-native collectives for purposes like protocol governance or pooled investment.
- **Web3:** A vision of the next-generation internet that is decentralised and user-controlled. Web3 services leverage blockchains for identity, data and transactions to reduce reliance on centralised platforms. It essentially envisages a "decentralised internet" built on blockchain foundations(McKinsey, 2023).
- **Tokenisation of Assets:** Converting rights in a real-world asset into a digital token on a blockchain. This links underlying assets to tradable tokens (Davis et al., 2024), enabling fractional ownership and peer-to-peer exchange of assets like equity, debt or real estate.
- **Decentralised Digital Identity:** Using blockchain to give individuals control over their own digital credentials (self-sovereign identity). Instead of relying on an authority, users hold verifiable identity attestations in a wallet and present them selectively. This lets people manage their identity data privately (Okta, 2024) while still enabling trusted credential verification.
- **Supply Chain & Material Passports:** Leveraging blockchain for traceability of goods and materials. Walmart found a blockchain system could trace food contamination in seconds rather than days; similarly, *material passports* record detailed data about products or building materials on blockchain to support recycling and circular economy goals (Shojaei and Naderi, 2024).

### 1.6.2 Blockchain Levels of Sophistication

Blockchain technology has evolved significantly since its inception, and it can be categorised into different levels of sophistication. These levels range from basic ledger systems to advanced platforms supporting complex applications.

### 1.6.3 New Service Layers

As blockchain technology continues to mature, new service layers are being developed to enhance its capabilities and address existing limitations, as shown in Table 1.2.

*Table 1.2* Blockchain Service Layers

| *Level* | *Core Technical Capability* | *Typical Innovations & Use-Cases* | *Indicative Gartner-Cycle Status (2025 → 2035)* | *Business Readiness* |
|---|---|---|---|---|
| 1 Distributed-Ledger Foundations | Shared, immutable ledger for timestamped transactions. (Narayanan et al., 2016). | Bitcoin payments; basic token transfers; audit trails. | *Slope of Enlightenment* (well-understood, commoditised) | Production-grade services widely available; low barrier to entry. |
| 2 Programmable Token Platforms | Smart contracts and native token issuance. (Buterin and Wood, 2013). | Asset tokenisation, ICOs/STOs, NFTs (digital art, IP). | Emerging from *Trough of Disillusionment* after NFT boom-and-bust. | Feasible with strong contract-security and legal opinion. |
| 3 Decentralised Service Layers | Composable Dapps; Layer-2 scaling (roll-ups, channels). (Wood, 2014). | Decentralised Finance (lending, AMMs); DAOs for protocol governance. | *Late Slope of Enlightenment*; stress-tested exploits. | Growing institutional pilots; risk-management standards maturing. |
| 4 Enterprise-Integrated Ecosystems | Cross-chain bridges, oracle networks, self-sovereign identity (SSI). | Supple-chain traceability, material passports, CBDC pilots, Web3 applications. | Between *Peak of Expectations* and early *Slope* (regulation/interoperability are bottlenecks). | Pilot to early production; consortium governance and standards work in progress. |
| 5 Autonomous Value Webs (≈ 2026–2030) | Machine-to-machine payments; zero-knowledge roll-ups; interoperable CBDCs. | IoT energy microgrids, real-time carbon credit clearing, industrial DAOs. | *Innovation Trigger* moving towards *Peak*. | Early pilots: mainstream uptake expected ~2030 when IoT and CBDCs converge. |

(*Continued*)

*Table 1.2* (Continued)

| *Level* | *Core Technical Capability* | *Typical Innovations & Use-Cases* | *Indicative Gartner-Cycle Status (2025 → 2035)* | *Business Readiness* |
|---|---|---|---|---|
| 6 Cognitive Decentralised Systems (≈ 2028–2033) | On-chain AI/ML operating on encrypted data; quantum-safe cryptography; self-optimising contracts. | Autonomous insurance pools, AI-managed trade finance, dynamic congestion-pricing tokens. | *Nascent Innovation Trigger* – research-stage. | Experimental sandboxes; regulatory frameworks for AI auditability pending. |
| 7 Ubiquitous Trust Fabric (≈ 2033–2035+) | Global, quantum-secure trust layer underpinning ever device, identity and contract; seamless cross-domain composability. | Borderless SSI backbone, universal device passports, programmable public-good funding. | Horizon view beyond current Gartner cycle; enters *Slope* post-2035. | Long-terms strategic planning; blockchain becomes invisible "plumbing" of the digital-physical economy. |

**Layer 2 Solutions** are designed to improve the scalability and transaction speed of existing blockchain networks. These solutions operate as secondary frameworks or protocols built on top of primary blockchains. For example, the Bitcoin Lightning Network allows for fast, off-chain transactions that are later settled on the main blockchain, significantly reducing congestion and transaction fees (Poon and Dryja, n.d.).

**Interoperability Protocols** play a crucial role in enabling different blockchain networks to communicate and interact with each other. These protocols facilitate a more connected and efficient blockchain ecosystem by allowing the seamless transfer of assets and data across multiple blockchains. Projects like Polkadot and Cosmos are at the forefront of developing interoperability standards that promote collaboration and innovation across the blockchain space (Zamyatin et al., 2019).

**Decentralised Finance (DeFi)** is one of the most transformative service layers built on blockchain technology. DeFi encompasses a wide range of financial applications that operate without traditional intermediaries such as banks. These applications include lending, borrowing, trading, and insurance services, all of which are accessible to anyone with an internet connection. DeFi platforms leverage smart contracts to provide transparent, efficient, and secure financial services, democratising access to financial markets and fostering financial inclusion.

## 1.7 Impact and Ramifications in the Built Environment

The Hackitt Report, officially titled the "Independent Review of Building Regulations and Fire Safety: Final Report," was authored by Dame Judith Hackitt and published in May 2018 (Hackitt, 2018). This comprehensive review was commissioned in response to the Grenfell Tower fire in June 2017, which tragically claimed 72 lives. The report examines the regulatory framework and related systems governing high-rise residential buildings and makes significant recommendations to improve safety and accountability.

The Hackitt Report uncovered significant flaws in the current building regulations and fire safety systems. It highlighted a lack of clarity and responsibility, with complex and fragmented regulations that created confusion among those tasked with ensuring building safety. The report also noted insufficient oversight and enforcement, with many buildings not adhering to existing regulations. Additionally, it criticised the prevalent "race to the bottom" mentality, where cost-cutting and poor-quality materials were prioritised over safety considerations.

To address these issues, the Hackitt Report proposed a complete overhaul of the regulatory system. Key recommendations included the establishment of a new regulatory framework that emphasises accountability and transparency, with clear roles and responsibilities for all stakeholders involved in building safety. The report also advocated for the creation of a new Joint

Competent Authority (JCA) to oversee the safety of high-rise residential buildings throughout their lifecycle. Furthermore, it called for a more robust and transparent product testing regime and stricter enforcement mechanisms to ensure compliance with safety standards.

The Hackitt Report highlighted the urgent need for a systemic overhaul of building regulations and fire safety to prevent tragedies like Grenfell from happening in the future. By addressing the shortcomings in the regulatory framework, clarifying responsibilities, improving competence, ensuring better product testing, and engaging residents, the report aimed to create a safer environment for occupants of high-rise residential buildings.

The "Golden Thread" is a central concept introduced in the Hackitt Report to ensure building safety throughout the lifecycle of high-rise residential buildings. This concept emphasises the importance of maintaining comprehensive, accurate, and up-to-date information about a building, from its design and construction through to its occupation and eventual demolition. The Golden Thread aims to enhance accountability, transparency, and communication among all parties involved in the building process.

## 1.8 Building Blockchain's Wider Ecosystem

Smart contracts are self-executing agreements automatically executing and enforcing predefined conditions without intermediaries, ensuring automation, immutability, and transparency. Once deployed, smart contracts cannot be altered, ensuring consistent enforcement of terms. This technology is revolutionising industries such as real estate, where ownership transfers can be automated, and supply chain management, where goods transfer can be verified without manual intervention (Buterin and Wood, 2013).

Smart contract can work hand to hand with Dapps, these are applications that run on a blockchain network instead of a centralised server, using smart contracts to perform functions and offer services without intermediaries. They benefit from decentralisation, incentivisation through tokens, and enhanced security due to their blockchain foundation. DApps are transforming sectors like finance, where DeFi platforms offer lending, borrowing, and trading without traditional banks, and gaming, where players can securely own and trade in-game assets (Wood, 2025).

With smart contract and Dapp, it is possible to create DAOs operating through rules encoded as smart contracts on a blockchain, managed by members via a decentralised voting system. They ensure transparency and trust through blockchain-based voting and automated operations, removing the need for centralised leadership. DAOs are reshaping governance models, enabling democratic decision-making in venture capital investments and social causes, where collective management of funds ensures transparency and fairness (Hassan and De Filippi, 2021).

**Interoperability and Cross-Chain Solutions:** Interoperability allows different blockchain networks to communicate and interact, facilitated by cross-chain solutions like atomic swaps and cross-chain bridges. These technologies enable seamless transfer of assets and data across blockchains, fostering innovation and collaboration. Projects like Polkadot and Cosmos are developing standards for cross-chain interactions, allowing DeFi platforms to integrate with various cryptocurrencies, enhancing user options and liquidity (Zamyatin et al., 2019).

**Layer 2 Solutions:** Layer 2 solutions are protocols built on top of existing blockchains to enhance scalability and efficiency by processing transactions off-chain. Technologies like state channels, Plasma, and roll-ups reduce congestion and transaction fees. These solutions are crucial for mainstream adoption, enabling fast and low-cost transactions in payment systems, like Bitcoin's Lightning Network, and reducing gas fees in DeFi platforms, thus improving user experience and platform efficiency (Poon and Dryja, 2015).

Chapter 2

# Blockchain Industrial Adoption

## 2.1 Initial Industry Reactions

Since the debut of blockchain technology in Satoshi Nakamoto's Bitcoin White Paper in 2008 (Nakamoto, 2008), there has been a significant progression from niche cryptocurrency experimentation to the broader, enterprise-level deployment of decentralised solutions. Blockchain's foundational attributes – namely, transparency, immutability, and trust minimization – have appealed to a variety of sectors, including to finance, healthcare, logistics, supply chain management, and government services (Casino et al., 2019; Zheng et al., 2018). Despite this growing interest, widespread adoption has been relatively restrained, often inhibited by regulatory uncertainty, integration challenges with legacy systems, and lack of clear business cases.

A systematic review by AlShamsi et al. (2022) reveals that studies investigating blockchain adoption frequently rely on well-established technology acceptance frameworks such as the Technology Acceptance Model (TAM) and the Technology–Organization–Environment (TOE) paradigm (Davis, 1989; Tornatzky et al., 1990). These analyses consistently identify trust, cost implications, perceived usefulness, and organisational support as critical drivers of acceptance. Barriers, on the other hand, include technological immaturity, privacy concerns, and interoperability hurdles (Zheng et al., 2018). Although supply chain management remains a dominant arena for empirical work, emerging use cases have appeared in diverse settings from tracking agricultural produce to automating complex financial instruments.

By 2025, market projections indicate a significant uptick in the scale and sophistication of blockchain-based services, spurred by growing regulatory clarity, standardisation efforts, and the convergence of blockchain with other Fourth Industrial Revolution technologies (e.g. the IoT, AI). Furthermore, organisations are experimenting with smart contracts to optimise transaction flows, while consortium-led blockchains are addressing interoperability challenges. Despite this progress, an under-representation of qualitative research suggests that key human, cultural, and organisational aspects often remain

DOI: 10.1201/9781003518303-3

under-explored, warranting further investigation into how decentralised networks reshape relationships between stakeholders.

Blockchain's journey from a cryptocurrency infrastructure to a transformative digital innovation underscore both its disruptive potential and its ongoing evolution. As organisations embrace more robust frameworks and align blockchain initiatives with broader strategic goals, the technology's capacity to yield transparent, efficient, and resilient processes will continue to expand.

## 2.2 Main Industrial Domains of Applications

Blockchain's versatility has led to its adoption in a range of industries, with supply chain management emerging as one of the most prevalent domains, as shown in Figure 2.1. Here, distributed ledgers enable real-time tracking of goods, improved provenance verification, and streamlined logistics.

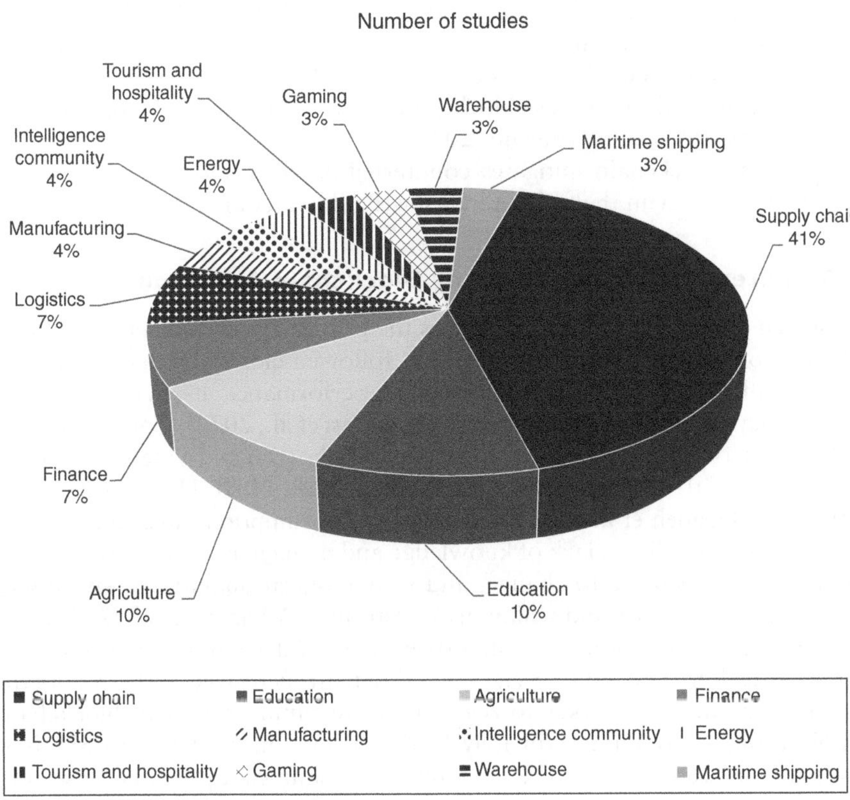


*Figure 2.1* Blockchain adoption by Industry.

Meanwhile, the financial services sector continues to pioneer novel blockchain applications, focusing on cross-border payments and asset tokenisation.

Healthcare organisations leverage blockchain to secure patient records, share data more efficiently, and ensure traceability of pharmaceutical products. Beyond these established fields, government services employ blockchain to enhance digital identity systems and increase transparency in public administration. Educational institutions are also experimenting with blockchain-driven certification processes to combat diploma fraud and facilitate secure credential transfers. Across these diverse settings, blockchain's core attributes of immutability, transparency, and trust minimisation underpin its growing relevance and potential impact.

Blockchain technology has found applications across diverse sectors:

- Supply Chain Management: Blockchain enhances traceability and transparency in supply chains, particularly in agriculture, logistics, and cold chain management (Vadgama and Tasca, 2021; Zhang et al., 2025).
- Finance: Decentralised finance platforms leverage blockchain for secure transactions and new business models (George Lawton, 2025).
- Healthcare: Blockchain facilitates secure sharing of patient records and reduces silos in healthcare systems (Reiff et al., 2024).
- Education: Universities use blockchain for credential verification and academic record-keeping (Rashid, 2024).
- Fashion: Blockchain mitigates counterfeiting risks and improves supply chain efficiency in the fashion industry(Badhwar et al., 2023).

## 2.3 External Factors Affecting Blockchain Adoption

Research consistently identifies trust as the primary factor influencing blockchain adoption, as shown in Figure 2.2, followed closely by perceived cost, social influence, facilitating conditions, performance expectancy, effort expectancy, and information security (AlShamsi et al., 2022). Conversely, significant barriers include security and privacy risks (Dai and Vasarhelyi, 2017; Treiblmaier, 2018) privacy risks (Esmaeilian et al., 2020; Hatzivasilis et al., 2021; Kouhizadeh et al., 2019) high energy consumption, investment costs, organisational policies, lack of knowledge and management support, cultural differences, reluctance to change, and technological immaturity—including interoperability issues and scalability constraints (AlShamsi et al., 2022).

Blockchain's adoption across industries strongly depends on understanding these critical drivers and barriers – notably trust, perceived cost, and organisational readiness. Market forecasts indicate rapid growth in blockchain applications – from approximately USD 800 million in 2017 to USD 64.4 billion by 2026 (Mehra, 2022) – highlighting significant potential in sectors like finance, healthcare, and supply chain management. Empirical research

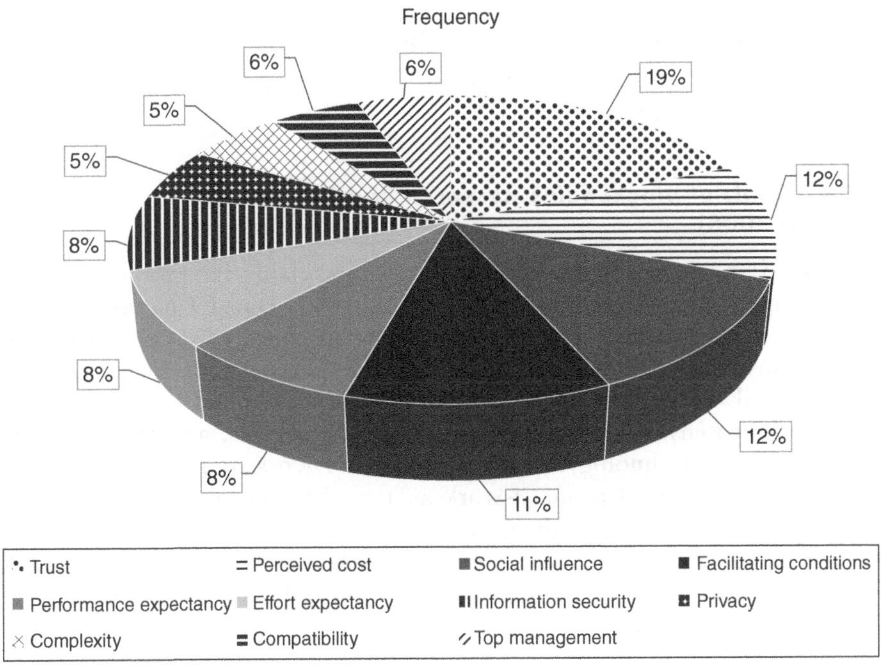


*Figure 2.2* External factors affecting Blockchain adoption.

frequently employs questionnaires (77% of studies) to investigate adoption factors, highlighting a preference for comparative quantitative analyses.

Supply chain management is frequently identified as a prominent sector for blockchain use, although banking and finance may hold similar prominence. Adoption frameworks across sectors consistently highlight trust, social influence, and perceived benefits as drivers, while high energy usage, investment expenses, cultural mismatches, and technological challenges such as interoperability continue to limit broader adoption.

Most existing studies (approximately 80%) focus on initial adoption (intention-to-use) rather than sustained or actual use, resulting in limited understanding of long-term deployment. Additionally, half of the reviewed studies emphasise top management perspectives, underscoring strategic-level decision-making over operational or user-level considerations. Methodological constraints such as heavy reliance on quantitative approaches and limited online data sources further narrow the scope. To better capture blockchain's cross-sector potential, future research should explore acceptance and ongoing usage behaviours, utilise diverse theoretical frameworks, adopt qualitative methodologies, and consider broader cultural and organisational contexts.

## 2.4 Blockchain Adoption Timeline

The trajectory of blockchain adoption in the industrial sector can be delineated into distinct phases.

### 2.4.1 *Initial Blockchains Research and Development (1982–2008)*

Prior to 2008, foundational research efforts laid the groundwork for Distributed Ledger Technology (DLT). In 1982, David Chaum proposed a vault system in his dissertation, introducing concepts such as achieving consensus among nodes and immutably timestamping data. In 1991, Stuart Haber and W. Scott Stornetta described a cryptographically secured chain of blocks to ensure the integrity of digital documents, an early precursor to blockchain technology. In 1993, Cynthia Dwork and Moni Naor proposed the concept of proof-of-work as a mechanism to deter spam emails and service abuses, laying the groundwork for later applications in securing distributed systems. These early research efforts collectively contributed to the development of DLT, culminating in the introduction of Bitcoin in 2008 (Sherman et al., 2019).

### 2.4.2 *Exploration and Experimentation (2008–2013) Emergence of Blockchain (Bitcoin's Early Days)*

In 2009, **Bitcoin** introduced the first public blockchain, demonstrating a peer-to-peer, decentralised ledger for digital currency. This "Blockchain 1.0" era was defined by cryptocurrency transactions and basic scripting, with blockchain technology largely synonymous with Bitcoin. Early milestones (e.g. the first real-world Bitcoin payment for pizza in 2010) proved the concept's viability but the ecosystem was volatile and unregulated. Mainstream businesses were initially wary – blockchain's Wild West reputation made it seem "too risky for business use," and few enterprises engaged beyond accepting Bitcoin for payments. By 2013, interest in cryptocurrency grew (Bitcoin's market cap exceeded $1 billion), yet blockchain applications remained limited to simple value transfer, and scalability and trust concerns kept most corporations on the sidelines.

### 2.4.3 *Early Adoption (2014–2015) – Smart Contracts and Early Enterprise Exploration*

Blockchain 2.0 emerged as innovators expanded blockchain's utility beyond digital cash. In late 2013, Ethereum's white paper proposed a platform for programmable smart contracts, and by 2015 the Ethereum "Frontier" network launched, allowing developers to write decentralised applications

(DApps) on a live public blockchain . This was a pivotal technological leap: Ethereum's smart contracts enabled complex business logic on-chain, opening use cases like asset issuance, automated agreements, and tokenisation. Around the same time, forward-looking enterprises began exploring blockchain tech in earnest. 2014 saw financial institutions shift focus "from digital currency to the development of blockchain technologies," and even consumer-facing firms like Microsoft, PayPal, and Expedia started accepting Bitcoin, validating broader interest. In 2015, the first enterprise blockchain consortia formed: the Linux Foundation launched the Hyperledger Project to develop open-source blockchain frameworks, and a group of nine major banks created the R3 consortium to research blockchain for banking, which grew to 40+ financial institutions within months. These initiatives signalled a new phase where businesses and tech leaders saw potential in private and permissioned ledgers for enterprise use, distinct from public cryptocurrencies.

### *2.4.4 Expansion and Integration (2016–2017) Enterprise Consortia and Private Blockchains*

By 2016, blockchain was gaining corporate credibility. An open-source community was actively developing enterprise-grade platforms, moving beyond the "risky" crypto image. Key developments in this period include J.P. Morgan's introduction of Quorum (2016) – a private, Ethereum-based blockchain tailored for business needs like privacy and high throughput. Industry consortia multiplied: the Enterprise Ethereum Alliance (EEA) formed in early 2017 with 30 founding corporate members (spanning banks, tech firms, and startups) to standardise Ethereum for business, and it swelled to over 150 members in six months. Likewise, regional and sector-specific consortium projects sprang up (for example, seven European banks launched the Digital Trade Chain in 2017 to trial blockchain in trade finance . On the technology front, permissioned blockchain frameworks matured – R3's Corda (focused on financial agreements) was unveiled to partners in 2016 and reached a 1.0 release by late 2017 after two years of development, and IBM and others contributed to Hyperledger's modules (e.g. Fabric and Sawtooth). By the end of 2017, blockchain fever had truly hit: Bitcoin's price surged to nearly $20k (drawing massive attention), and about *15% of global banks* reported using blockchain in some capacity . This era saw blockchain utility extend into enterprise pilots and prototypes, with private/consortium blockchains addressing business requirements for permissioned access, privacy, and performance that public blockchains couldn't yet meet.

### *2.4.5 From Pilots to Production Deployments (2018–2019)*

**Pilot projects** from prior years began evolving into real deployments, marking a leap in blockchain maturity for business. In mid-2017 Hyperledger

Fabric (an open-source permissioned ledger) reached **version 1.0**, its first production-ready release . This enabled companies to move from proof-of-concept to building real solutions on a stable platform. By 2018, enterprise blockchain applications were launching across industries: IBM and Maersk introduced **TradeLens (2018)**, a global shipping ledger aimed at streamlining supply chains, and IBM's Food Trust network went live to trace produce for retailers like Walmart. In fact, **Walmart rolled out a Hyperledger-based food traceability system in 2019**, mandating suppliers to use blockchain for tracking produce. Financial services also progressed beyond experiments. J.P. Morgan led a landmark institutional adoption with the 2019 debut of **JPM Coin**, a dollar-backed token for interbank payments – the first cryptocurrency issued by a major U.S. bank, meant to enable instant settlement using blockchain. Major tech providers further legitimised enterprise blockchain by offering managed services: **Amazon's AWS launched Amazon Managed Blockchain in 2019**, supporting Hyperledger and Ethereum to help businesses easily create blockchain networks in the cloud . Meanwhile, public blockchain scalability and utility improved, narrowing the gap for enterprise use. New high-performance chains and layer-2 solutions emerged (e.g. EOS launched in 2017 promising commercial-grade DApp support, and the Bitcoin Lightning Network saw growing adoption for faster transactions). By 2019, the Ethereum network was handling **over 1 million transactions per day**, demonstrating capacity growth. In summary, this period transformed blockchain from experimental pilots into **early production systems**. Enterprises achieved initial ROI in areas like supply chain tracking, trade finance, and cross-border payments, even as they continued to grapple with integration challenges and industry coordination.

## 2.5 Scaling and Broader Institutional Adoption (2020–2021)

Entering the new decade, blockchain began integrating into core business strategies. A 2020 Deloitte survey found nearly **40% of global executives had blockchain in production** and over half viewed it as a top-five strategic priority – a testament to how far enterprise perception had come. This maturation was facilitated by improvements in scalability and tooling. In late 2020, Ethereum initiated its **Eth2 upgrade (Beacon Chain)** as a first step towards proof-of-stake and sharding, aiming to massively increase throughput and reduce fees. The focus on scalability extended to consortium platforms as well, with performance tuning and interoperability becoming key goals (e.g. Hyperledger Fabric 2.0 and Corda enhancements). **Blockchain-as-a-Service** offerings multiplied; by 2021, all major cloud providers (AWS, Azure, IBM, Oracle) offered blockchain managed services, lowering adoption barriers. Crucially, **institutional and corporate adoption accelerated**. In 2020, enterprise Ethereum got a boost as ConsenSys acquired JPMorgan's Quorum platform, consolidating enterprise Ethereum development. Large enterprises started leveraging both private and public blockchains: for example, **PayPal and Visa entered the crypto space** (PayPal enabling crypto purchases in 2020, and in 2021 Visa even began settling transactions using USDC stablecoin on Ethereum). The **decentralised finance (DeFi)** boom of 2020–21, while centred on public networks, spurred financial institutions to pay attention – by late 2021, DeFi services on Ethereum reached $200B in value, hinting at new models for lending, trading, and liquidity that banks began cautiously exploring. **NFTs** also exploded in 2021, with NFT artwork fetching millions at Christie's auction. This NFT craze pulled consumer brands and media companies into the blockchain arena to issue digital collectables, further normalising the technology. Meanwhile, the cryptocurrency market hit a $3 trillion capitalisation in 2021, and Coinbase's high-profile public listing on NASDAQ signalled mainstream acceptance of the crypto industry. Collectively, these milestones indicate that by 2021, blockchain had moved beyond pilot projects into a phase of **scaling and institutional embrace** – both through dedicated enterprise blockchain platforms and increased interaction with public blockchain ecosystems.

### *2.5.1 Consolidation, Caution, and Convergence (2022–Present)*

The past few years have seen blockchain reach a more **sober maturity** in business settings, marked by both consolidation of efforts and convergence between private and public networks. The year 2022 tested the industry's resilience: crypto markets experienced a sharp downturn (with events like the Terra stablecoin collapse and FTX exchange failure), which tempered the prior hype. Some high-profile enterprise blockchain initiatives struggled or shut down, illustrating practical challenges. Notably, **IBM and Maersk's TradeLens** (Cecere, n.d.) **shipping blockchain was discontinued in 2022 due to insufficient network adoption**, despite onboarding nearly 100 participants. Similarly, the Australian Securities Exchange cancelled its multi-year project to replace its settlement system with a blockchain implementation, citing complexities. These setbacks taught hard lessons about scaling consortia and the importance of industry-wide buy-in. Many enterprises regrouped, narrowing their blockchain projects to well-defined, value-add use cases and improving interoperability with existing systems.

At the same time, **enterprise blockchain continued to advance and integrate**. A significant milestone was the **Ethereum "Merge" in 2022**, where Ethereum transitioned to proof-of-stake consensus – reducing energy usage by ~99% and addressing environmental concerns that had made some companies hesitant to use public blockchains. The emergence of more energy-efficient, scalable chains (and Layer 2 networks) aligned blockchain tech with corporate ESG goals and performance requirements. By 2022, NFTs and digital assets remained in focus (many brands launched NFT collections or metaverse experiments), and more *"eco-friendly blockchain networks emerged"* to support these activities.

Crucially, the **line between permissioned and public blockchain began blurring.** In a groundbreaking pilot in late 2022, **JPMorgan executed a DeFi trade on a public blockchain (Polygon)** as part of a regulated experiment, swapping tokenised currencies using smart contracts. This demonstrated that with proper safeguards, even conservative financial institutions can utilise permissionless networks for efficiency gains. Many central banks and governments also moved from research to implementation phases for digital currencies: by the end of 2022, over **100 countries were exploring or developing Central Bank Digital Currencies (CBDCs)** – effectively government-backed, permissioned blockchain systems for fiat money. This trend represents institutional adoption at a national scale, aiming to leverage blockchain for faster, more transparent monetary transactions.

By 2023, enterprises proceeded more cautiously but still *"moving ahead with blockchain"*, focusing on practical applications in finance, supply chain, gaming, real estate, and healthcare. Regulatory clarity slowly improved (e.g. frameworks like the EU's Markets in Crypto-Assets (MiCA) were introduced), further encouraging institutional participation. There is also a pivot towards **"Web3"** paradigms – many businesses see potential in blockchain's role in a decentralised internet for data security, digital identity, and direct user ownership. The convergence of blockchain with other emerging tech is accelerating integrating **AI, IoT, and smart contracts** to automate and secure processes (for example, using IoT devices with blockchain for trusted supply chain tracking).

As of 2025, the enterprise blockchain landscape is more **measured and robust.** Companies have learned which use cases benefit most from blockchain's strengths (such as provenance tracking, inter-company data sharing, and digital asset settlement) and are less inclined to use blockchain as a solution in search of a problem. Permissioned blockchains (like Hyperledger Fabric, R3 Corda, and ConsenSys Quorum) are deployed in production consortia for trade finance, insurance, and logistics, albeit sometimes with smaller scale than originally hyped. Public blockchain networks (like Ethereum and its Layer 2 extensions) are increasingly viewed as **global shared infrastructures** that businesses can tap into for certain tasks – for instance, to leverage the security and liquidity of a public network when appropriate, while keeping sensitive data on private channels. The focus now is on improving **interoperability and scalability** to handle enterprise volumes: sidechains, cross-chain bridges, and standardised protocols are being developed so that private and public networks can securely interact.

## 2.6 Conclusion

As blockchain technology continues to mature, its adoption across various industries is poised for significant growth. Projections indicate that the global blockchain technology market, valued at approximately USD 31.28 billion in 2024, is expected to reach USD 1,431.54 billion by 2030, reflecting a compound annual growth rate (CAGR) of 90.1%. This rapid expansion is driven by increasing demand for decentralised services and continuous advancements in blockchain protocols.

One of the critical challenges hindering widespread blockchain adoption has been scalability and interoperability. Innovative solutions, such as Layer 2 scaling technologies and cross-chain protocols, are being developed to address these issues, enabling higher transaction throughput and seamless communication between disparate blockchain networks. These advancements are anticipated to facilitate more efficient and integrated blockchain applications across various sectors.

The convergence of blockchain with emerging technologies like the IoT and AI is also expected to unlock new possibilities. For instance, the integration of blockchain with IoT is projected to enhance security and accelerate digital transactions, with the blockchain IoT market expected to reach $12,679.5 million by 2030, growing at a CAGR of 58.2%. Similarly, AI integration can lead to more intelligent and autonomous systems, further driving blockchain adoption.

In the financial sector, the adoption of blockchain is accelerating, with traditional financial institutions integrating blockchain technology into their services. For example, PayPal has introduced its proprietary stablecoin for transactions, and Visa has launched the Visa Tokenised Asset Platform (VTAP) to connect fiat currencies with blockchain, enabling banks to handle fiat-backed tokens and stablecoins. These developments indicate a growing acceptance and integration of blockchain in mainstream financial services.

As regulatory frameworks become more defined, industries are likely to gain greater confidence in implementing blockchain solutions. The European Union's implementation of the MiCA regulation in 2023 provides a comprehensive regulatory framework for cryptocurrencies and blockchain applications within member states, signifying a move towards clearer guidelines and support for blockchain innovation.

In conclusion, the maturation of blockchain technology, coupled with the development of scalable and interoperable solutions, is set to drive its widespread adoption across various industries. The integration with emerging technologies and the establishment of supportive regulatory frameworks are paving the way for more seamless and innovative industrial applications, positioning blockchain as a foundational pillar for transparent, efficient, and collaborative environments.

Chapter 3

# Blockchain in the Construction Sector

Until 2015, blockchain was largely confined to cryptocurrencies, and its application in construction was non-existent. The foundational concept of an immutable, shared ledger for distributed computing was established, and by 2015 the advent of Ethereum enabled programmable smart contracts, setting the stage for industry-specific use cases (Cheng et al., 2021). Forward-looking firms and academics began speculating how these innovations could improve construction workflows (e.g. automating contract execution or securing project data), but no significant on-site trials occurred yet. By 2020 a growing awareness of blockchain's potential led to the first wave of pilot projects and consortia in construction. Global think-tanks like the World Economic Forum (WEF) started promoting blockchain for supply chain resilience. By the late 2010s, industry consortia and R&D initiatives were exploring blockchain-based solutions for construction supply chains and contract management. Early pilots were typically small-scale proofs-of-concept aimed at demonstrating feasibility. For example, by 2019, prototypes showed how BIM models could be linked with smart contracts to trigger payments upon milestone completion (Pham et al., 2025). In academia, researchers identified five key benefit areas for blockchain in AEC industries (Cheng et al., 2021), including:

- Supply chain management,
- Contract and payment management,
- Information management,
- Stakeholder management, and
- Integration management.

However, overall adoption remained nascent and fragmented, with construction "still lagging behind other industries" in blockchain usage.

In the 2020s, commercial implementations in construction began to emerge, especially in regions with strong digitalisation agendas. Several pilot projects matured into operational systems, validating efficiency gains. By the end of 2020, for instance, China had deployed blockchain at scale

DOI: 10.1201/9781003518303-4

in construction: over 160 construction projects used blockchain-based platforms to pay 40,000+ migrant workers, eliminating intermediaries and ensuring on-time wages (Cai et al., 2021). Such successes, along with accelerating digital transformation due to the COVID19 pandemic, spurred global interest. Major infrastructure owners and governments in Asia and Europe launched blockchain initiatives for procurement, permitting, and asset tracking. By 2025, blockchain and smart contracts are recognised as part of the construction innovation toolkit worldwide. The WEF even projects that 10% of global GDP could be stored on blockchains by 2027, across industries (Allaire, 2024) – underscoring the significant role construction may also play in that blockchain-based economy. Still, adoption is uneven globally, as detailed in the regional sections of this chapter.

## 3.1 Construction Case Studies by Geographic Regions

### *3.1.1 UK*

The UK has been a pioneer in exploring blockchain for construction, driven by a combination of industry consortium efforts and government-backed innovation programs. Early on, UK-based engineering firms and academics formed collaborative groups to investigate use cases. For example, the *Construction Blockchain Consortium* (CBC) was established in 2018 at University College London as an industry-led hub to prototype blockchain solutions and share knowledge across firms (Smith, 2019). Around the same time, the *Construction Smart Contract Committee* – comprising major companies like Arup, SAP, IBM, and Tata Steel – began developing common standards and pilot projects for smart contracts in construction. These partnerships aimed to mitigate the risk of early adoption by pooling expertise and resources on pilot trials.

#### *Supply Chain Traceability*

A notable UK pilot was launched by Tata Steel in 2018, in collaboration with SAP, IBM and Arup, to create a *"transparent chain of custody"* for construction materials (Smith, 2019). This pilot tracked steel components from sourcing through installation using a blockchain ledger, ensuring that material certifications and quality inspections were immutably recorded. The motivation was to improve sustainability in procurement – Tata Steel's report had identified that knowing material provenance (e.g. recycled content and carbon footprint) is crucial for lifecycle sustainability. By logging all material data on blockchain, stakeholders could instantly verify quality and origin, increasing confidence in the supply chain. Early results indicated that such a system could streamline compliance with specifications and make it easier to reuse or recycle materials at end-of-life.

### *Smart Contracts for Payments and Project Delivery*

The UK government's Industrial Strategy Challenge Fund supported pioneering trials of blockchain in project management using Smart Contracts for Payments and Project Delivery. One high-profile project is "PLASMA," a two-year (2018–2020) collaboration between Vinci Construction UK, Skanska UK, and partners like the Building Research Establishment (BRE) and tech firm nPlan (Construction Management, 2019). Funded by UK Research and Innovation, PLASMA set out to develop a blockchain-based digital platform for construction planning, progress tracking, and automated payments. The proof-of-concept, built on a private blockchain, enables planners to rehearse project sequences and then record site events (deliveries, task completions, etc.) as tamper-proof data blocks IoT sensors were integrated to automatically log task completion, which in turn could trigger *"incremental payments"* via smart contracts tied to milestones. This approach promised to reduce delays in payment processing and enhance trust among clients, main contractors, and subcontractors. Vinci and Skanska reported that, if scaled, the tool could deliver 25% savings in cost and time on projects by optimising schedules and eliminating paperwork. While PLASMA was a proof-of-concept, its development was a key milestone demonstrating real-world feasibility of blockchain in UK construction. The *success* of such pilots has been cited by UK officials – for instance, the head of the UKRI Transforming Construction program noted that projects like PLASMA *"play a key role in advancing transformation across the sector"*.

### *BIM Integration and Asset Management*

The UK has also explored linking blockchain with its national BIM efforts. Building Information Modelling was mandated for public projects (Level 2) by 2016, and the vision for BIM Level 3 is a fully collaborative 3D model shared by all parties. Security and data trust are vital in that context. A 2018 report by Arup argued that blockchain could "dramatically improve transparency and security of collaboration in BIM" (Smith, 2019). Currently, BIM data is exchanged via centralised common data environments (CDEs), which pose risks of single-point failure or unauthorised modifications. Blockchain, by contrast, can maintain an immutable, timestamped log of all BIM model changes. In theory, this guarantees that the *as-built* digital twin remains a trusted record for the asset's life. Major UK contractors have internally evaluated such concepts – envisioning, for example, a system where IoT sensors on prefab components update the BIM model and automatically trigger payment release via smart contract once installation is verified (Smith, 2019). Although as of 2025 fully blockchain-enabled BIM Level 3 is not yet realised, these ideas have influenced ongoing R&D and standards discussions in the UK. The BSI and institutions like the Institution of Civil Engineers (ICE) have published guidance (e.g. an ICE Insights report in 2019) encouraging firms to run *"small-scale proof of concept"* trials to validate blockchain benefits before wider rollout.

Overall, the UK's early forays have shown tangible benefits in niche applications: improved supply chain visibility, faster payment cycles, and more secure collaboration on project data. The presence of active networks (Maciel, 2025) and government innovation funding has positioned the UK as a leading testbed. By 2025, many UK construction organisations have gained at least pilot experience with blockchain, and best practices (for example, data standards for smart contracts) are starting to be codified to support broader adoption. The UK experience also highlighted challenges – notably the need for cultural change and clarity in legal frameworks – but it demonstrated that blockchain *can* work alongside construction IT systems to deliver efficiency and trust.

## *EU*

Across the EU, interest in blockchain for construction has been driven by initiatives around cross-border collaboration, regulatory compliance, and sustainability. The EU's focus on the digital single market and emerging regulations (like electronic procurement and waste tracking) created a supportive environment for blockchain experimentation. While the EU did not see as many large-scale construction pilots as the UK or China, several noteworthy projects and policy-driven trials took place. Many of these were funded through research programs or conducted by multinational consortia, reflecting the EU's emphasis on standardisation and interoperability.

### *Material Traceability and Circular Economy*

A key use case in Europe is using blockchain to track construction products for sustainability (recycling, carbon reporting, etc.). For example, a European tech firm Compellio partnered with EU authorities in a 2023 *Blockchain Sandbox* program to pilot a solution for *circular supply chains in the construction sector* (Develle, 2024). The system logs product data (origin, composition, recyclability) on a ledger, addressing privacy and scalability needs through a permissioned network. The anticipated benefit is increased trust and verifiability of product sustainability claims, enabling easier compliance with upcoming EU regulations on construction product transparency. By storing certifications and workflow steps immutably, such a platform helps auditors and clients ensure that, for instance, a given batch of cement or steel meets the promised environmental standards throughout its lifecycle. This kind of pilot aligns with the EU's broader circular economy goals and could streamline compliance with new waste and recycling standards in construction.

### *Cross-Border Digital Procurement*

The EU's single market involves complex logistics and paperwork when building materials move across countries. Blockchain has been tested to digitise and simplify these processes. A pioneering example was the ACCUDIRE pilot, developed in Italy, 2021 – regarded as Europe's first fully digital international shipment using blockchain. In this pilot, a shipment from Italy to Turkey had all its logistics documents (e.g. e-CMR consignment notes) digitised and secured on a blockchain platform (Develle and Thomas, 2024). Participants included a major manufacturer, port authorities, customs, a freight forwarder, and the importer, all interfacing via a shared ledger. Digital signatures (with geolocation) of the sender, carrier, and receiver were recorded on the blockchain to provide an immutable, traceable record of the cargo's journey. The blockchain interoperability with customs' systems allowed automatic pre-notification of goods and real-time verification of document authenticity. This *"Pilot 1.0"* demonstrated that even complex, multi-party construction logistics (in this case, it was cited in EU forums as a template for paperless transport) can be streamlined – eliminating couriered papers, preventing loss of documents, and speeding up customs clearance. The success in Italy has been referenced in EU discussions (DTLF and eFTI initiatives) as a model for scaling paperless trade in construction supply chains across member states.

### *Integration With BIM and Digital Twins*

Europe's strong BIM community has also looked at blockchain as an enabler for data sharing. EU-funded research (Horizon 2020 and others) produced frameworks for *smart construction contracts using BIM and blockchain*, showing how a BIM model could validate progress and then a smart contract automate the corresponding payment (Pham et al., 2025). Pilot implementations of these ideas have been limited, but one notable case is from Finland, which combined regulatory change with blockchain technology to digitise an aspect of construction: property transactions.

### *Blockchain-Powered Home Sale*

In 2019, Finland saw one of the world's first fully digital real estate transactions on blockchain. A homebuyer in Helsinki purchased a new apartment from builder Skanska entirely online – without a single sheet of paper (Skanska, 2019). This pilot was enabled by a new Finnish law creating a digital apartment ownership registry (in lieu of traditional paper certificates), which took effect in January 2019. Skanska teamed with a tech startup to develop a blockchain platform that records the purchase contract, signatures, and ownership transfer in a secure, immutable manner. The buyer, Liisa Lintula, became a pioneer showcased in media for executing a housing transaction through a few clicks, backed by blockchain's *"single source of truth*"(Skanska, 2019). This project demonstrated the seamless integration of construction handover with blockchain – the moment the building was ready for sale, all documentation was handled digitally, drastically reducing the time and friction in the sales process. The benefits observed include saved time (no need for in-person meetings or mailing documents) and enhanced security (fraudulent title transfers are virtually impossible with the blockchain-based registry). While this is a real estate use case, it directly ties into the construction lifecycle (final stage of delivering a unit to a buyer) and showcases how smart contracts could one day manage not just construction workflows but also post-construction transactions. The Finnish system is now an early template that other EU countries are studying as they modernise land registries.

In summary, the EU's engagement with blockchain in construction has been characterised by strategic pilots that align with regulatory innovation. The region has leveraged blockchain especially for improving transparency and efficiency in cross-organisational processes – whether tracing materials for sustainability or exchanging data across borders. Many efforts are still in pilot or research phases as of 2025, but they have laid groundwork for broader adoption. The anticipated benefits in the EU context include easier compliance with EU-wide directives (through trusted data), reduction of administrative costs in transnational projects, and greater assurance of sustainability metrics on projects. As EU institutions continue to fund blockchain projects and develop common standards, further implementations in public procurement and infrastructure projects are likely in the coming years.

### *United States (USA)*

In the United States, the adoption of blockchain and smart contracts in the construction industry has been gradual and led primarily by private sector innovation rather than government mandates. Throughout the 2010s, the US construction sector observed blockchain developments abroad and in other industries (like finance or supply chain) with interest, but only a handful of early adopters undertook pilots. By the early 2020s, awareness had grown significantly, and construction firms – especially large contractors and tech startups – began experimenting with blockchain solutions targeting persistent industry pain points like payment delays and document management. The regulatory environment in the US has been somewhat cautious (no specific federal programs for blockchain in construction), so most progress came from company-led initiatives and collaborations.

#### *Startup-Driven Pilot Projects*

Several construction-tech startups in the US emerged to bring blockchain to construction. One example is Brickschain, a California-based startup founded in 2018 specifically to apply blockchain in construction management. In a notable pilot, Brickschain partnered with Probuild, one of Australia's largest construction firms, to implement a blockchain system for tracking construction assets and materials across the supply chain. The project pulled data via API from various suppliers' systems (spanning China, Hong Kong, and Australia) and recorded the provenance of each building component – from factory to site – on a shared ledger. Even site facilities like temporary offices and equipment were tagged and tracked. Upon project completion, the building owner received a comprehensive digital ledger of every installed element, with documentation of supplier, manufacture date, and warranties, all secured on blockchain (Morris, 2018). This pilot demonstrated improved supply chain transparency and facilities management: owners could easily verify the origin and specs of a component years later by querying the blockchain record, which is far more reliable than paper trails. Probuild's managing director noted that having *"the blockchain as the source of truth"* and integrating it with IoT (Bluetooth Low Energy tags on shipments) provided live tracking and confidence in data integrity. He also hinted at future steps like integrating a banking partner to automate financing and payments based on the secured data flow. Brickschain's success with Probuild led it to secure partnerships involving construction revenues over $20 billion, indicating that several large contractors (including in the US) were testing its solutions (Morris, 2018). This case exemplifies how US startups have been key catalysts, often proving the technology overseas (where project owners were willing to experiment) and then aiming to bring it back to the US market.

### *Smart Contract Payment Trials*

The chronic issue of slow and complex payment processes in US construction has made automated payment smart contracts an attractive concept. Some general contractors and software firms have piloted systems on a limited basis. A typical scenario involves linking project management software or on-site sensors with a smart contract: for example, when a building's construction reaches 25% completion (as verified by a sensor or an inspector's digital sign-off), the smart contract automatically releases the next payment to the contractor (ConExpo, 2025). Such a workflow was described in a 2023 industry report, highlighting how milestone-based automatic payments could eliminate paperwork and delays. Early trials on US projects have used IoT devices (drones, cameras, or wearables) to monitor progress and then trigger payment events on blockchain networks (often using private Ethereum-based ledgers for confidentiality). While specific company names are not always public, it is known that major US construction software providers (like Procore and Oracle Aconex) have explored integrating blockchain modules. The *potential benefits* demonstrated in these trials are significant: faster payments to subcontractors (improving cash flow for small trades) and reduced disputes, since the contract terms and evidence of work are transparent to all parties on the ledger. A faster payment cycle also reduces the need for expensive financing or liens. However, uptake in the US has been slow – many firms treat these as experimental. By 2025, smart contract payment systems are not yet mainstream in the US, but proof-of-concept projects have shown clear advantages in efficiency and trust. Industry forums like the Associated General Contractors (AGC) and CONEXPO have featured these successes to educate the market.

### *Industry Awareness and Indirect Support*

Though no federal construction blockchain programs exist, the US has seen supportive developments. The insurance and surety sector, for instance, is looking at blockchain to manage surety bonds and insurance certificates on projects (which could later dovetail with construction smart contracts for risk management). Moreover, some state-level initiatives intersect with construction – e.g. state pilots to put *land titles* or *building permits* on blockchain (Ohio and Illinois had early pilots for land registries around 2019, and New York explored blockchain for municipal record-keeping). These indirectly build comfort with the technology in the real estate and construction ecosystem. By the mid-2020s, *awareness* in the US construction industry has grown to the point that many large firms have internal teams or consultants assessing blockchain's readiness. Trade publications frequently discuss how smart contracts can expedite project delivery and reduce admin overhead. In fact, surveys of construction executives in 2024 showed a much higher familiarity with terms like "blockchain", even if active implementation lags behind Europe or Asia. The general sentiment is that blockchain is "just beginning" in construction but holds clear potential if challenges (legal uncertainties, interoperability with legacy systems, etc.) can be overcome.

By 2025, US has moved from theoretical discussion to initial practice in niche areas of construction. Notable benefits observed in pilots include real-time supply chain visibility, improved payment speed, and more secure record-keeping (e.g. immutable as-built data). However, broad adoption is still on the horizon – the *lessons learned* from pilots are helping craft the business case needed to convince the traditionally conservative US construction industry. As more case studies demonstrate ROI (return on investment), it is anticipated that the US will see an uptick in commercial implementations, possibly accelerated by the entrance of enterprise tech giants (IBM, Oracle, etc.) offering blockchain-enabled construction management tools.

## ***Asia-Pacific***

The Asia-Pacific region has been a hotbed of blockchain adoption in construction, led by strong government support in some countries and a general culture of rapid tech adoption. Notably, East Asia and the Middle East have implemented blockchain in construction at scales not yet seen elsewhere, often as part of national digitalisation or smart city programs. Key themes in Asia's adoption include ensuring integrity of public works (e.g. preventing corruption in payments), improving labour welfare, and linking blockchain with other advanced technologies like IoT and digital twins in mega-projects. We highlight developments in several leading countries:

### *Large-Scale Implementation in China*

China's central government embraced blockchain as a strategic technology around 2019 (with President Xi Jinping publicly endorsing it), and the construction sector became one of the focus areas. By the end of 2020, China had integrated blockchain-based systems in major public construction projects, particularly to manage project finances and workforce payments. A flagship example is the blockchain platform in Xiong'an New Area (a massive smart city project in Hebei province). In Xiong'an, the local government and the state-owned ICBC bank jointly deployed a blockchain system to handle compensation and wage payments on construction sites. This was done to reduce intermediaries and ensure funds reach the intended recipients without delay or misappropriation. The impact has been remarkable: *"by the end of 2020, more than 160 construction projects and 40,000 migrant workers were paid via blockchain systems"*, according to official reports. This addresses a long-standing issue in China where migrant construction workers often faced wage arrears. With smart contracts automating the release of payments upon work completion (and all transactions recorded immutably), trust between contractors and labourers improved and wage fraud became significantly harder. In addition to payments, Chinese authorities have used blockchain for materials quality tracking and project approvals. For instance, some provinces require that key construction quality tests (like concrete strength reports) be uploaded to a blockchain ledger that regulators can audit in real-time, ensuring data cannot be falsified. By 2022, China also released numerous policy documents (217+ by various ministries) on blockchain integration across industries, including construction standards. Tech giants like Tencent and Alibaba have enterprise blockchain divisions that collaborated with construction firms to build platforms – e.g. Tencent's TrustSQL was used in a pilot to trace the supply chain of rebar steel on a Shenzhen high-rise project (improving detection of counterfeit materials). Overall, China's approach has been top-down: mandate trials in government projects, demonstrate benefits (efficiency, reduced corruption), and then encourage wider industry adoption. The observed benefits include faster processing of administrative tasks, greater data sharing among agencies (since blockchain acts as a common data backbone), and increased confidence in the integrity of construction records. China's example shows that with sufficient political will, blockchain can be scaled up for construction – although it's worth noting many implementations use permissioned (private) blockchains due to the need for controlled access and the enormous transaction volumes in large projects.

### *Automated Payments in Singapura's Public Infrastructure*

Singapore is another leader, known for its smart city initiatives and high-tech construction practices. In 2024, Singapore made headlines by piloting a blockchain-based payment system for public construction projects. The *Land Transport Authority (LTA)*, which oversees major infrastructure like highways and metro lines, partnered with OCBC Bank to implement a conditional payments solution on blockchain (Sin, 2024a). Specifically, this was applied to the new Cross Island MRT Line contracts. Under the pilot, LTA disburses mobilisation advances (the upfront payments given to contractors to kick-start work) through a smart contract platform integrated with the banking system. The smart contract is coded with the conditions of the advance – for example, it can release funds in tranches only when certain proof-of-expenses or progress is logged by the contractor. As a result, LTA gets full transparency on how the advance is utilised by the contractor via a real-time dashboard, and funds cannot be diverted to unrelated purposes. By November 2024, the pilot with three main contractors had successfully disbursed over S$22 million through this system. The benefits reported include faster processing of payments (straight-through, with automatic verification instead of manual checks) and improved governance – essentially eliminating the risk of misuse of advance payments and providing an auditable trail of every dollar. This is critical for public accountability in multi-billion dollar projects. Singapore's example integrates blockchain with traditional banking, showing a hybrid model where a permissioned blockchain handles the logic and transparency, while actual funds move in the banking system once conditions are met. Given the success, Singapore is looking to extend this to other payment types (e.g. stage payments to subcontractors) and it serves as a model for other Asian nations looking to digitise public works procurement.

### *Supply Chain and Modular Construction in Australia*

Australia, part of the Asia-Pacific sphere, has also conducted interesting pilots. The earlier mentioned Probuild-Brickschain project is one example of a cross-continental blockchain supply chain application, connecting Chinese suppliers to an Australian jobsite. Additionally, Australian construction firms have explored blockchain to improve prefabrication and modular construction workflows. Researchers in Hong Kong and Australia collaborated on a *blockchain-enabled digital twin platform* for modular construction, where each module's data (from fabrication to on-site assembly) is recorded on blockchain, enabling all stakeholders to see the status and trust the updates. A *real-life pilot* of such a cyber-physical system in construction showed it could reduce information latency and errors in coordination (Xiao et al., 2023). While these studies are in early stages, they signal the region's interest in combining Digital Twins (real-time digital replicas of physical construction) with blockchain for accountability. In practical terms, this could mean a crane lifting a prefab module triggers an IoT sensor that logs the event on blockchain, updating the digital twin and notifying all parties (engineers, site managers, insurers) instantly in a verifiable way. The benefit is a highly synchronised construction process with minimal disputes about what happened when – since the ledger provides a shared, trusted timeline.

### *Government-Led Innovation in the Middle East (Gulf)*

In the Middle East, the United Arab Emirates (UAE), especially Dubai, has pursued a government-driven blockchain agenda. Dubai's "Blockchain Strategy 2020" aimed to move all government transactions to blockchain, which includes areas like building permits, land registries, and contracts. By 2019, Dubai Land Department had introduced a blockchain system for real estate transactions, and there were pilot programs for digital building permit approvals wherein multiple approval agencies (planning, utilities, civil defence) share a blockchain to sign off permits sequentially. This has *anticipated benefits* of reducing approval times from weeks to days and eliminating forgery of permits. Saudi Arabia and Qatar have also investigated blockchain for managing their mega-projects (like NEOM in Saudi or 2022 FIFA World Cup stadium projects in Qatar) to ensure secure project management and procurement. While detailed results from these were not widely published, there is evidence that Gulf mega-projects have tested blockchain to oversee complex supply chains (for instance, tracking imported construction materials to ensure they meet specifications and are sustainably sourced, as part of their commitment to quality).

Across Asia-Pacific, the impacts observed include greater trust in multiparty collaborations (often an issue in large projects with many subcontractors), reduction in fraud or errors (especially in financial transactions and material certifications), and improved efficiency in administrative processes. Notably, Asia has shown that blockchain can tie into social objectives in construction: China's case improved migrant workers' livelihoods by guaranteeing timely pay, and other countries have considered blockchain to enforce safety compliance (e.g. a worker's training certifications on a blockchain to control site access). The anticipated benefits in Asia are thus not only economic (time/cost savings) but also social and governance-related – increased transparency to the public for government spending on infrastructure, assurance of ethical practices in construction (preventing use of conflict materials or unfair labour), and creation of trusted "digital twins" of built assets that will serve smart cities in the long term. With Asia-Pacific governments and corporations continuing to invest in this technology, the region is expected to remain at the forefront of blockchain-in-construction adoption.

### *Wider Cross-Regional Initiatives*

Beyond individual regions, there have been global efforts and collaborations shaping the adoption of blockchain in construction. One such initiative is the CBC (originating in the UK but with international members) which regularly hosts conferences and publishes open research to disseminate learnings globally. Similarly, organisations like buildingSMART International (which standardises BIM data formats) have started task groups to explore how blockchain could complement existing standards for data exchange in construction. The aim is to ensure that as blockchain solutions develop, they remain interoperable across borders – a crucial factor for global construction firms operating in multiple countries.

The WEF has been an influential voice in advocating blockchain for supply chains in general. In 2019, the WEF released *"Building Block(chain)s for a Better Planet,"* highlighting how blockchain could improve transparency in industries including construction (for example, by tracing raw materials like timber to ensure they come from legal, sustainable sources) (Allaire, 2024). In 2020, as mentioned, WEF launched a Blockchain Deployment Toolkit to guide companies through implementing pilot projects (Nelson, 2021). Construction companies, including global engineering firms, participated in the WEF's pilot programs, sharing case studies on early blockchain use. This global knowledge-sharing has helped spread successful use cases (a company in one country might replicate a proven pilot from another).

Another cross-regional development is the work on standards for smart contracts and legal recognition. Organisations such as the International Federation of Consulting Engineers (FIDIC) and various national construction law associations have been discussing how contract forms (e.g. standard contracts for construction projects) might evolve to accommodate or explicitly allow smart contract mechanisms. Some governments (like the UK Law Commission and Singapore's Smart Nation initiative) have issued clarifications on the legality of smart contracts, which gives confidence to international projects that code-based contracts will be enforceable.

From a technological standpoint, open-source platforms have emerged that target construction: for instance, Hyperledger Fabric, an open-source blockchain framework, has been utilised in many of the projects across different regions (UK's PLASMA, some EU pilots, and parts of China's implementations all chose permissioned Hyperledger networks). This means improvements or modules built for one project can often be reused by another – an important factor accelerating adoption globally.

By 2025, globally operating construction and engineering firms (e.g. Skanska, Vinci, ACCIONA, Samsung Engineering, etc.) have started to include blockchain considerations in their digital strategy. Many have run internal proof-of-concepts across multiple offices/regions to evaluate benefits uniformly. This cross-regional perspective is vital, for example, to ensure that a supply chain blockchain used in Europe can hand off data to an Asian client's system, or that a digital twin on blockchain in one country can be understood by facilities managers in another country where the asset is located.

Overall, the global momentum suggests that blockchain and smart contracts are transitioning from a buzzword to a practical tool in construction. Importantly, the global community is learning that one size does not fit all – the optimal approach in a high-trust environment like Scandinavia may differ from that in a low-trust environment where blockchain's tamper-proof features are most needed. International collaboration helps identify which use cases have universal value (supply chain provenance and payment automation appear near-universal) and which require local tailoring (such as legal smart-contract enforcement or integration with country-specific procurement platforms).

### Observed Impact and Benefits

After a decade and a half of development, the impacts of blockchain and smart contracts on the construction sector are becoming clearer through these pilot projects and implementations:

#### *Transparency and Trust*

Virtually all successful use cases report improved transparency among stakeholders. Whether it's knowing exactly where a batch of materials came from or seeing a real-time ledger of payments, blockchain provides a *"single source of truth"* that all parties can trust. This has led to reduced disputes – for example, automatic milestone payments leave little room for quarrel over who approved what and when, since the conditions and outcomes are recorded immutably. In multi-tier supply chains, trust is boosted: subcontractors are more willing to work with unfamiliar general contractors if payments are guaranteed by a smart contract escrow, and owners have more confidence that contractors won't cut corners if all critical data (tests, inspections, etc.) must go on the chain. This **shift from paper-based trust to digital trust** could gradually improve the adversarial culture that has long plagued construction.

#### *Efficiency and Cost Savings*

Many pilots, such as the UK's PLASMA and Singapore's LTA project, quantified potential efficiency gains. Removing manual reconciliation of documents and automating processes can save significant labour hours and shorten project delivery times. The *25% cost and time reduction* targeted by the PLASMA toolkit is illustrative of these high aspirations. While not every pilot will achieve such numbers at scale, even single-digit percentage improvements on multi-billion-dollar projects translate to substantial savings. Moreover, real-time data sharing (e.g. site IoT devices feeding a blockchain) means project managers can make quicker, better-informed decisions, potentially reducing costly delays. Early evidence from China and elsewhere indicates that administrative costs (related to audits, compliance, contract management) can be trimmed when a trusted ledger is in place – for example, Xiong'an's inter-departmental blockchain linkage eliminated many duplicate data entry tasks between agencies, enhancing efficiency of oversight (Cai et al., 2021).

### *Security and Data Integrity*

Construction projects generate massive amounts of documentation. Storing this in a tamper-proof way has clear benefits for risk management. Several case studies (like the BIM timestamping in the UK and digital permit systems in Dubai) show that blockchain can prevent unauthorised modifications or loss of critical records (Smith, 2019). For owners and insurers, this means audit trails of construction quality are available decades later (useful in disputes or maintenance). Also, cybersecurity is enhanced – rather than trusting emailed PDFs or printed certificates (which can be forged), stakeholders rely on cryptographically secured records. However, it is also noted that blockchain is not a panacea for all security issues – the *"whole system"* (from sensors to user interfaces) must be secured, as one UK report cautioned. Still, on balance, blockchain's introduction has increased the overall integrity of construction data management, especially when combined with technologies like digital signatures and IoT-based verification.

### *Stakeholder Satisfaction and Fairness*

A less tangible but important benefit observed is improved stakeholder satisfaction. Contractors and subcontractors getting paid on time (or even faster, in smaller increments) via smart contracts report better cash flow and reduced financial stress. Clients benefit from greater visibility into project progress and fund usage, which improves client-contractor relationships. In environments where corruption or late payments were rampant, blockchain has acted as an *"equaliser,"* enforcing rules uniformly. For instance, the Chinese worker wage platform essentially enforces labour law in code, ensuring every worker is paid promptly for logged work – a clear social benefit. Similarly, automated compliance checks (like ensuring only certified materials are used) create a fairer market for suppliers (cheating is harder, so honest suppliers aren't undercut by fraudulent ones). These improvements, while hard to quantify, contribute to a healthier construction ecosystem and could make the industry more attractive to investors and new talent.

Looking ahead, many anticipated benefits are expected to materialise as pilot projects scale up. Integration with Digital Twins is poised to unlock new value – for example, a building's digital twin on blockchain could continuously log performance data (from sensors) throughout the building life, creating an immutable maintenance history that boosts resale value and safety. Smart cities may use blockchain to integrate construction data with city operations, enabling, say, automated issuance of completion certificates once a smart contract confirms all required inspections are done. The combination of blockchain with emerging tech like AI (for predictive analytics on project data) and 5G (for instantaneous data capture on-site) suggests that the construction site of the future could be highly autonomous and self-verifying.

Nevertheless, some challenges temper the full realisation of these benefits. Interoperability between different blockchain systems is an ongoing issue – efforts like China's Blockchain-based Service Network (BSN) and the EU's EBSI aim to address this by creating common infrastructure (Cai et al., 2021; Ekman, 2021). Legal and contractual frameworks need to catch up; for instance, who is liable if a smart contract makes an incorrect payment due to a sensor error? These questions are being examined by academics and lawyers (as referenced in many systematic reviews (Cheng et al., 2021), and preliminary guidance exists, but real legal precedents will likely be set in the coming years as disputes (inevitably) arise in blockchain-managed projects.

## 3.2 Blockchain-in-Construction Enterprises

Construction projects involve complex, multi-party co-dependent processes often mired in inefficiencies – long supply chains and fragmented communication lead to disputes, delays, and cost overruns. For CEOs and directors, the challenge is leveraging blockchain strategically to address these pain points while aligning with enterprise principles like long-term strategic fit, organisational readiness, and technological integration.

### *3.2.1 Strategic Alignment with Business Goals*

Adopting blockchain should begin with clear strategic alignment to enterprise goals. Leaders must identify where blockchain adds value – for example, improving trust in supply chains, automating payment processes, or enhancing compliance. The technology's core strengths (an immutable, shared ledger) directly address the industry pain points by increasing transparency and accountability across stakeholders. Self-executing smart contracts can automate complex contractual workflows, reducing administrative friction and speeding up project delivery. In one Harvard Business Review analysis, a global construction adviser noted that automating contract processes via blockchain "could save money, free up valuable resources, and speed up project delivery" (Tapscott and Vargas, 2019). Similarly, consulting studies predict significant gains: a PwC report estimates up to *30% faster project delivery* from improved transparency and less paperwork. To ensure strategic fit, executives should tie blockchain initiatives to key objectives such as cost reduction, risk management, and client satisfaction. Notably, 40% of construction executives in a Deloitte survey believe blockchain will be "critical" to their operations within five years – underscoring that aligning projects with long-term business strategy is imperative for staying competitive (A3Logics, 2025).

### 3.2.2 *Organisational Transformation and Governance*

Implementing blockchain is not just a technology upgrade; it demands organisational transformation. Construction enterprises must foster a culture of collaboration and innovation to realise blockchain's benefits. New governance structures may be needed to manage shared ledgers that span multiple firms. Leaders should evaluate which processes to standardise and share on a blockchain network versus where to continue competing. As one industry expert observed, construction firms are learning to *"think carefully about when and where we need to compete and what we can share and collaborate on"* (Tapscott and Vargas, 2019). Blockchain consortia or partnerships can define common data standards and trust frameworks so that contractors, subcontractors, clients, and regulators participate with confidence. Change management is critical: staff and partners require training on blockchain platforms and new workflows (e.g. handling digital contracts and wallets). Early pilot projects can help demonstrate value to stakeholders and build buy-in. Organisational policies must also adapt – for instance, integrating blockchain controls into procurement, project management, and compliance processes. Critically, robust governance and legal frameworks should accompany any blockchain rollout to address liability, data privacy, and regulatory compliance (Kun, 2021). Construction executives should treat blockchain as a catalyst to re-engineer processes (like payments, document approvals, supply tracking) in a more transparent, efficient mold, rather than a plug-and-play IT solution.

### 3.2.3 *IT System Integration and Enterprise Architecture*

From an enterprise architecture perspective, blockchain solutions should be designed as an integrated layer within the existing IT landscape of the construction firm. Rather than a standalone system, blockchain must interface with project management software, ERP systems, BIM databases, and other digital tools. A guiding principle is to avoid data silos – the blockchain network should pull in and push out data through APIs so that all stakeholders see a "single source of truth" across systems. For example, California-based startup *Briq* built a blockchain platform that integrates with existing workflows and systems to chronicle the entire building process, creating an immutable record of project data. By linking with familiar software (scheduling, accounting, BIM), such solutions minimise disruption and ensure users can access blockchain benefits (e.g. tamper-proof records, automated verifications) within their normal tools. Enterprise architects should also consider permissioned blockchain networks which restrict access to authorised parties – a crucial feature for industry consortia handling sensitive project data. Security and interoperability standards must be defined at the outset so that the blockchain layer aligns with the enterprise's security architecture and data governance policies. The experience of a Singapore pilot is instructive: a major bank and transit authority co-developed a blockchain payment system integrated into the bank's platform, giving real-time visibility of transactions while automating contractor payments once preset conditions were met. This illustrates how thoughtful integration can embed blockchain into core business processes (in this case, payment disbursements) without overhauling existing IT infrastructure. In summary, treating blockchain as an extension of the enterprise architecture – subject to the same rigour of system design, cybersecurity, and interoperability – will ensure it bolsters rather than disrupts the overall IT ecosystem.

## 3.3 Blockchain-in-Construction Projects

The effect of blockchain in reshaping the management of construction projects if particularly noticeable in the architecting of CDEs in respect to project management structures. These environments integrate various digital processes, such as BIM, enterprise content management, site surveys, and reality capture, etc.. Blockchain provides stakeholders with real-time, verified project and design data, establishing a reliable foundation for single source of truth. By securely tracking model revisions, task completions, and contractual changes, blockchain significantly reduces disputes and streamlines automated functions like payments and compliance validations. As participants enter or leave projects, blockchain maintains continuity through a transparent, shared repository, facilitating informed, data-driven decisions. This integration makes BIM workflows increasingly collaborative, also transforming CDEs into robust, trusted platforms that improve overall construction management efficiency.

### *3.3.1 Supply Chain Management*

Blockchain can provide end-to-end visibility and traceability of construction materials, from sourcing to delivery on-site. This can help to verify the authenticity and sustainability of materials, track their provenance, and ensure timely delivery, reducing delays and improving overall project efficiency. For example, a blockchain-based system could track the journey of steel from the manufacturer to the construction site, recording quality checks, certifications, and delivery times at each stage. This level of transparency can help to build trust among supply chain partners and mitigate risks associated with counterfeit or substandard materials, aligning with ethical and quality considerations.

### *3.3.2 Transparency and Dispute Resolution*

The immutable and transparent nature of blockchain can significantly reduce disputes among project stakeholders. All key project information, including contracts, change orders, inspection reports, and payment records, is securely and transparently recorded on the blockchain, providing a single, auditable source of truth. In the event of a disagreement, the blockchain record can serve as irrefutable evidence, facilitating faster and more efficient resolution of disputes, saving time and costs associated with lengthy legal processes.

### *3.3.3 Compliance and Regulation*

Blockchain can help to streamline compliance with building regulations and other statutory requirements4 .... Records of approvals, certifications, and inspections stored on the blockchain can provide irrefutable proof of compliance to regulatory authorities. Smart contracts can also be programmed to automatically enforce compliance checks at various stages of the project, ensuring that all work adheres to the required standards. This can reduce the risk of penalties and delays associated with non-compliance.

## 3.4 Conclusion

From 2009 to 2025, the journey of blockchain and smart contracts in the construction sector has moved steadily from hype to reality. Early optimism in the mid-2010s – when blockchain was often touted as "the next IT revolution" for construction – has given way to a more nuanced understanding of where these technologies truly add value (Smith, 2019). Globally, dozens of pilot projects have tested blockchain across various construction use cases, and a growing number of commercial implementations are now live, especially in Asia and parts of Europe. Supply chain traceability, automated payments, digital twins, and BIM integration have all been proven on a pilot scale. The benefits observed include stronger trust, reduced delays and disputes, better accountability, and some early evidence of cost savings – all aligning with long-standing goals for improving construction productivity and collaboration.

Regional analysis reveals a diverse landscape: the UK spearheaded early trials and consortium-based innovation, linking blockchain with BIM and procurement improvements (CM, 2019; Smith, 2019). The EU leveraged pilots to aid sustainability and cross-border processes, preparing the ground for standardising these approaches (Develle, 2024). The USA, while slower to adopt, has seen startup-driven use cases that highlight the technology's potential in the American context (e.g. supply chain verification and faster subcontractor payments). Asia-Pacific demonstrates how government backing can propel blockchain into mainstream construction workflows – from China's massive deployment ensuring wage fairness to Singapore's use of bank-integrated smart contracts for public projects (Sin, 2024b).

It is evident that blockchain is not a silver bullet for all construction challenges – many pilots encountered hurdles such as the need for digital literacy among staff, integration with legacy systems, and unclear legal recognition. Some in the industry remain sceptical, as construction is traditionally cautious with new technologies. However, the trend from 2009 to 2025 is one of increasing maturity: early scepticism is gradually being overcome by solid demonstrations of value. The coming years will likely see broader rollout of successful pilot solutions (for example, nationwide adoption in China of the blockchain wage system, or multiple EU countries adopting material traceability networks), and the entrance of large enterprise vendors will make blockchain solutions more accessible to mid-sized construction firms.

The adoption of blockchain and smart contracts in construction by 2025, though uneven, has firmly shown promising improvements in project delivery and management. Many pilot projects have transitioned into real deployments, yielding benefits in supply chain efficiency, payment speed, data security, and stakeholder trust. As academic and industry knowledge continues to grow, and as standards and best practices emerge, we can anticipate a construction industry that increasingly leverages blockchain for a more transparent, collaborative, and reliable project ecosystem. The foundational work done from 2009–2025 globally will serve as the springboard for construction's next evolution – one where digital and physical processes are interlinked via blockchain to deliver better outcomes for all participants.

Chapter 4

# Blockchain Oracles in Construction

## 4.1 Key Definitions

Oracles are key technological components that can significantly enhance blockchain-based solutions for the AECO sector. As construction projects continue to generate vast amounts of data – from sensor readings on-site to regulatory requirements – ensuring the seamless flow of information between on-chain and off-chain systems is of paramount importance. Oracles serve as trusted intermediaries, bridging real-world data with blockchain networks, thereby enabling intelligent smart contracts that can automate complex project tasks.

AI algorithms extract insights from project data to enable predictive analytics, risk management, and proactive decision-making across the project life cycle. These technologies also underpin automated compliance and quality control processes, helping to ensure that projects meet contractual, regulatory, and environmental requirements. By integrating oracles and AI, including ML, the AEC sector can realise greater transparency, improved asset traceability, and more efficient resource allocation, ultimately avoiding waste and driving sustainable development. This chapter explores what these technologies are and how they can be effectively applied. It also explores the challenges they pose, and the opportunities they offer to revolutionise construction data management, improve project outcomes, and support the transition towards Construction 4.0 and beyond.

## 4.2 Kinds of Oracles

### *4.2.1 Understanding the Oracle Concept*

Blockchain technology, by design, operates within a closed, deterministic environment, meaning that the execution of smart contracts relies solely on data already recorded on the chain. However, real-world applications in the AECO sector inevitably require data from external sources, such as weather updates, supplier databases, or government registries. Oracles function as

DOI: 10.1201/9781003518303-5

trusted data channels bridging these off-chain and on-chain environments. Essentially, they feed validated, real-world information into the blockchain so that smart contracts can trigger outputs based on dynamic, real-time conditions.

### 4.2.2 *Types of Oracles*

1. **Hardware Oracles**: These oracles gather data directly from physical devices such as sensors, drones, and radio-frequency identification (RFID) scanners (FDA, 2024). Hardware oracles can feed information about air quality, temperature, or vibration levels on a building site, thereby enabling smart contracts to enforce health and safety protocols automatically. Hardware oracles must be tamper-resistant, especially in high-risk environments, to ensure the accuracy of the data they collect.
2. **Software Oracles**: These oracles interface with software-based sources, including application programming interfaces (APIs), web services, and third-party databases (Mukhopadhyay, 2018). Examples include retrieving exchange rates for cross-border payments, automatically pulling compliance regulations from government portals, or obtaining BIM model updates from a shared project management platform (Hijazi et al., 2022).

   Software oracles rely heavily on the reliability of the data source. If a government database is not regularly updated or is compromised, the oracle can unwittingly transmit flawed information to the blockchain.
3. **Inbound vs Outbound Oracles**: **Inbound** oracles funnel external data into the blockchain—for instance, a weather monitoring service that signals a potential storm, prompting a construction site to cease certain operations. **Outbound Oracles** allow the blockchain to send data or commands to external systems—for example, once a payment milestone is verified on-chain, an outbound oracle can signal an enterprise resource planning (ERP) system to release funds or purchase new materials.
4. **Consensus-Based Oracles:** These oracles gather data from multiple sources and apply consensus mechanisms to reduce the risk of single-point failure or data manipulation (Chainlink, 2020). For high-stakes decisions—such as verifying the structural integrity of a critical component—consensus-based oracles can aggregate sensor data from multiple devices and cross-reference it with engineering standards before updating the blockchain.

### 4.2.3 *Security, Reliability, and Future Directions*

- **Security**: Oracles represent a potential point of vulnerability in any blockchain ecosystem (Mason and Escott, 2018). For the AECO sector, it is crucial to implement cryptographic authentication and redundancy measures to ensure data integrity.

- **Reliability**: Service-level agreements (SLAs) and reputation metrics can incentivise oracle providers to maintain high standards of data quality.
- **Future Directions**: As 5G networks and the Internet of Things (IoT) continue to mature, hardware oracles will become increasingly vital for real-time, high-frequency data flows in smart construction projects (Marinelli et al., 2023).

## 4.3 Bridging Real-World Data and Blockchain

### 4.3.1 *The Necessity of Bridging Mechanisms*

To realise the promise of blockchain in construction—enhanced transparency, immutable record-keeping, and automated project management—reliable data from the physical environment must inform on-chain decisions (Tapscott and Tapscott, 2016). Without a mechanism to bring off-chain data on-chain and vice versa, many real-world use cases would remain unachievable (Azhar et al., 2008).

### 4.3.2 *Architectural Considerations for Oracles in Construction Projects*

- **Integration with Existing Systems**: Project management tools (e.g., Primavera P6, Asta Powerproject) and BIM software (e.g., Autodesk Revit) already contain rich datasets. Oracles must be designed to integrate seamlessly with these platforms (Grilo and Jardim-Goncalves, 2010).
- **Data Standardisation**: The AECO sector lacks universal data standards, making interoperability a challenge (ISO 19650; Valero et al., 2022). Establishing common taxonomies and formats can facilitate smoother oracle implementation.
- **Scalability and Performance**: Complex construction sites generate vast volumes of data. Implementing real-time oracles at scale requires robust network infrastructure capable of handling large data streams without sacrificing security (Radman et al., 2025).

### 4.3.3 *Case Study: Automated Supply Chain Management*

Consider a major infrastructure project involving multiple subcontractors, each responsible for procuring specific materials. By using oracles linked to supplier APIs, the blockchain can monitor inventory levels and delivery schedules in real-time. Smart contracts can trigger payments automatically once deliveries reach the site and pass a quick quality check (Benton et al., 2018). This level of automation reduces administrative overheads, minimises delays, and ensures a transparent audit trail for all materials used in the project.

## 4.4 Conclusion

Oracles represent an emerging, yet potentially transformative, technology for bridging real-world data and blockchain ecosystems within the AECO sector. Their ability to reliably deliver off-chain data—ranging from sensor readings to regulatory updates—enables smart contracts to automate complex tasks, offering the possibility of enhanced transparency, greater efficiency, and improved compliance across construction projects.

However, realising these benefits demands careful consideration of security, reliability, and scalability. Hardware, software, inbound, outbound, and consensus-based oracles each come with distinct advantages and vulnerabilities, necessitating robust cryptographic safeguards and rigorous data quality assurances. Additionally, seamless integration with existing systems and industry-wide data standardisation are critical to overcoming interoperability challenges.

As 5G and IoT technologies evolve, hardware oracles are likely to play an increasingly central role in real-time, data-driven construction management. While current applications remain limited, early pilot projects – such as automated supply chain management – already highlight the considerable promise of oracle-enabled solutions. Moving forward, the AECO sector's proactive engagement with these innovations will be essential to fully harness their potential, driving the industry towards a more connected, sustainable, and productive Construction 4.0 future.

Chapter 5

# DAOs and Artificial Intelligence (AI) in Construction

## 5.1 An Introduction to DAOs

### *5.1.1 Core Definitions and Features*

A Decentralised Autonomous Organisation (DAO) is a self-governing entity that encodes rules and operational logic into a distributed ledger, commonly via smart contracts. Unlike traditional hierarchical firms, which depend on executives and managers to set priorities and sign-off decisions, DAOs place decision authority directly in the hands of their token-holding community. This authority can manifest in multiple ways – for example, through on-chain voting procedures that automatically trigger smart contract-based transactions if a proposal passes. In other words, the organisational "charter" is expressed by code, rather than reliant on continuous human oversight.

Ziegler and Welpe define DAOs as "a system in which storage and transaction of value and notary (voting) functions can be designed, organised, recorded, and archived; and where data and actions are recorded and autonomously executed in a decentralised way, typically requiring a blockchain or similar technology for consensus" (Ziegler and Welpe, 2022). Key aspects include:

1. **Decentralised Governance:** Decisions are taken collectively, often weighted by governance tokens or reputation. This ensures no single authority can unilaterally override the will of the community.
2. **Autonomy via Smart Contracts:** Pre-defined code automatically enforces decisions – e.g., releasing funds from a treasury or updating membership rights – upon approval.
3. **Shared Ownership and Transparency:** All financial and governance transactions are recorded on a blockchain, where they remain visible to anyone with access to the public ledger.
4. **Token-based Incentives:** Tokenomics align user behaviour with the DAO's objectives, such that members may be rewarded in tokens for their contributions, thereby fostering sustained collaboration.

DOI: 10.1201/9781003518303-6

### 5.1.2 DAO Creation Platforms and Software Services

Although pioneering DAO projects once required custom-written smart contracts and extensive technical knowledge, several software platforms now "offer DAO creation as a service," allowing novices to spin up new decentralised organisations with minimal code. Well-known examples include:

- **Aragon:** Provides modular DAO templates, governance plugins, and user-friendly interfaces. Commonly used for quickly establishing a governance token, voting mechanisms, and treasury management.
- **DAOstack (Alchemy):** Offers frameworks for holographic consensus, where users can propose initiatives and stake tokens to boost proposals they believe merit higher visibility.
- **Moloch DAO Frameworks:** Designed for smaller groups or membership-based DAOs; emphasises simplicity and straightforward on-chain voting with shares that can be "rage quit" at any time.
- **Colony:** Specialises in "task-based" DAO operations, assigning reputation to participants as they complete tasks. This approach aims to replicate dynamic project teams, especially relevant in complex project environments such as construction.
- **Snapshot (Off-chain Voting Portal):** While not strictly a DAO creation platform, Snapshot underpins many DAOs with lightweight, off-chain voting that references a blockchain snapshot of token balances.

For complex engineering and construction applications, platforms that allow flexible role definition, budget control, and milestone-based disbursements are particularly pertinent. Users can thus choose the framework that best fits their project's size, risk profile, and collaboration needs.

### 5.1.3 Visualising DAO Activity and Governance

Beyond set-up, DAO communities often rely on tools to visualise data, such as membership changes, treasury composition, proposal activity, and voting patterns. Open-source projects and analytics dashboards (e.g., DeepDAO, Dune Analytics) can aggregate on-chain data into dynamic charts. For instance, a construction-focused DAO might want to track how frequently certain project phases receive funding proposals or examine how many participants from specific trade disciplines (e.g., civil engineers versus architects) are actively voting. By plotting these metrics, stakeholders can:

- Pinpoint points of friction (e.g., repetitive proposal vetoes),
- Identify potential skill or funding gaps in the community,
- Increase accountability, by showing exactly who participated in or abstained from critical decisions,

- Demonstrate project progress to external parties, such as regulators or finance partners.

Such visualisations can also highlight trends in token holdings, making it easier to discern whether voting power is distributed equitably or if "whales" (members with large token balances) dominate key decisions.

## 5.2 Managing Construction Projects with DAOs

### 5.2.1 *Potential Benefits and Rationale*

Construction enterprises and project consortia frequently manage complex supply chains with multiple contractors, subcontractors, and consultants. This fragmentation often leads to siloed information flows, delayed payments, and contractual disputes. Because of its automated natures embedded in the blockchain, DAOs offer a new governance model that harnesses blockchain's transparency, potentially addressing enduring inefficiencies.

1. **Efficient Payment and Accountability:** Smart contracts can automate invoice settlement upon verified completion of tasks. For instance, if a subcontractor completes a pre-defined milestone, a proposal to release payment can be automatically enforced once on-chain approval is reached.
2. **Reduced Overheads:** Rather than relying on a project manager to gather votes or chase signatories, a DAO's contract-based processes keep track of deadlines, budget lines, and resource allocations. This reduces administrative burdens, saving time and money.
3. **Greater Trust Among Stakeholders:** In a traditional scenario, smaller contractors may doubt that prime contractors will pay them promptly, or that design changes won't cause hidden cost overruns. DAOs allow near real-time visibility into the status of budgets, schedules, and sub-treasury allocations, instilling confidence.
4. **Immutable Record-Keeping:** Contracts, approvals, and change orders recorded on-chain reduce confusion and disagreements about whether or when a decision was made.

### 5.2.2 *Practical Workflow Integration*

Integrating DAO structures into an existing construction project typically starts with a "hybrid" arrangement. For instance, a project governance board is formed with representatives from architecture, engineering, construction, and client teams. These representatives collectively set up a DAO treasury to handle specific budget lines—e.g., materials procurement. The project's standard contractual relationships remain in place, but the DAO approach covers:

- **Proposal Mechanisms for Budget Allocation:** Each discipline (e.g., mechanical, electrical, plumbing) can request funds for specific tasks. These requests appear on a DAO dashboard, and participants holding governance tokens or staked roles vote on whether to release payment.
- **Milestone-based Audits:** External auditors or quantity surveyors might be given restricted DAO membership to confirm that milestones are complete, or that a supplier has delivered correct materials. Their vote or signature then triggers the disbursement to the supplier.
- **Transparent Resource Tracking:** Because all transactions are recorded on the blockchain, managers can see precisely how each micro-budget is being spent, minimising the risk of duplicated or fraudulent invoices.

### 5.2.3 *Addressing Governance Challenges*

While theoretically appealing, construction-centred DAOs must confront organisational and cultural issues. Adapting project management norms—especially in large, joint-venture environments with strict legal frameworks—requires a careful design of the DAO's governance parameters:

- **Defining Voting Rights:** Should each stakeholder's voting power correspond to project share, the amount of capital invested, or the relative impact of their role (e.g., the structural engineer's sign-off may have greater criticality than that of a landscaper)?
- **Off-chain vs. On-chain Execution:** Many construction tasks still require in-person verification. Consequently, full on-chain automation might be restricted to financial disbursements or project updates, whereas physical inspections remain offline.
- **Legal Contracts:** Although DAOs operate trustlessly in code, real-world construction continues to rely heavily on legal contracts. Entities must clarify how to reconcile blockchain-based "smart" agreements with local building regulations, contract laws, and dispute-resolution processes.

Overcoming these issues demands stakeholder education, pilot programmes, and possibly incremental transitions from a conventional project management office to a partially on-chain environment.

## 5.3 The Role AI and ML in DAOs

### 5.3.1 *Overview: Synergy Between Decentralisation and AI*

AI and ML capabilities can be integrated into DAO ecosystems to optimise decisions, especially where large data sets are involved. In the construction sector, and with the advent of BIM, abundant data arise from design phases, procurement, scheduling, and on-site progress measurements. By leveraging immutable, on-chain records as reliable, tamper-proof inputs, AI-driven models can make more accurate projections and flag potential issues earlier than conventional systems.

Key advantages of combining AI with DAOs include:

- **Data Reliability**: Since DAO transactions and records are time-stamped and unalterable, ML pipelines can ingest more trustworthy data, reducing the risk of inaccurate inputs.
- **Decentralised Decision Execution**: Once a model flags a risk or identifies an optimisation opportunity (e.g., bulk ordering of supplies at a discount), the DAO community can vote, and smart contracts can swiftly implement approved actions.
- **Consensus on Models**: DAO stakeholders can collectively decide which AI models or data streams to trust, fostering transparency about the algorithms' assumptions and performance metrics.

### *5.3.2 Data Integrity and Oracles*

In many real-world use cases—particularly in construction—external data (e.g., site temperatures, material deliveries, local building code updates) must be fed into the DAO via "oracles." These are mechanisms that validate and transmit off-chain data onto the blockchain. By cryptographically signing data at the source, oracles help ensure its authenticity. AI engines may then process these inputs, offering predictive insights:

- **Progress Assessments**: For instance, comparing actual on-site progress (verified by IoT sensors or drone imagery) against planned schedules can highlight potential delays well before they become critical.
- **Resource Allocation**: An AI-based approach can adjust resource allocation dynamically, recommending how many workers or how much machinery to deploy, based on real-time constraints.

This synergy between decentralised ledgers and AI-driven analysis encourages trustless collaboration, allowing all parties to rely on a single, transparent source of truth.

### *5.3.3 Automation and Trustless Collaboration*

One of the most promising aspects of AI-integrated DAOs is autonomous decision execution. If a predictive model determines that a project is trending towards cost overruns, it can automatically draft a proposal in the DAO to reallocate contingency funds. If the voting passes, smart contracts release the required budget immediately. This process:

- **Reduces Administrative Delays:** Human project managers no longer need to relay routine decisions up and down a hierarchy.
- **Strengthens Accountability:** The algorithm's recommendations, together with the voting record, remain on-chain, enabling detailed audits and post-project analysis.
- **Facilitates Rapid Response:** In emergencies (e.g., unforeseen structural anomalies), an AI-based system can rapidly detect outliers and raise an immediate proposal to mobilise resources.

Of course, trusting critical decisions to algorithms requires careful calibration. Construction is inherently complex, and not all scenarios can be captured by code or statistics. Hybrid human–AI oversight remains advisable, ensuring that decisions with significant safety or financial repercussions receive due diligence from experienced professionals.

## 5.4 Predictive Analytics and Risk Management

### *5.4.1 Forecasting Techniques and Cost Control*

Machine learning-driven forecasting in construction typically applies regression or classification models to predict costs, timelines, and quality outcomes. By tying these techniques to a DAO's on-chain records, predictive models can be enhanced in several ways:

1. **Consistent, Verified Data**
   All project transactions—whether purchase orders or progress payments—are stored on a decentralised ledger. This builds a rich historical dataset that models can mine to anticipate overruns, cost spikes, or scheduling bottlenecks.
2. **Unified Data Sources**
   Often, construction projects store data in departmental silos. DAOs encourage shared data by design, meaning more comprehensive datasets for training ML algorithms.
3. **Data-Driven Decisions**
   Once the ML model flags a risk (e.g., a likelihood of steel shortages), the DAO community can propose an immediate strategy, such as placing orders earlier or allocating extra funds.

### *5.4.2 Risk Management in Multi-Party Projects*

Risk in large construction projects is distributed across multiple parties, including contractors, subcontractors, and suppliers. Because a DAO formalises collective governance, it can also embed risk analysis into its proposal processes:

- **Automated Alerts**: If on-chain data indicates that a subcontractor's tasks are consistently behind schedule, an ML model can forecast future delays or cost overruns. The DAO can then hold a vote to mitigate, whether by reallocating resources or switching suppliers.
- **Capital Reserves and Contingency**: Tokens in a DAO treasury can be earmarked for contingency budgets, only accessible if certain triggers are met—e.g., an AI risk rating surpassing a threshold that indicates likely project disruption.

These mechanisms allow more agile risk response, diminishing the chaos that arises when contractual or managerial silos delay timely interventions.

## 5.5 Automated Compliance and Quality Control

### *5.5.1 Smart Contracts and Regulatory Frameworks*

In many jurisdictions, construction projects must adhere to stringent building codes, health and safety legislation, and environmental standards. DAOs can assist compliance in several ways:

- **Programmed Milestones**: A smart contract could require specific regulatory sign-offs before releasing funds for subsequent phases, ensuring no step is overlooked.
- **Immutable Evidence**: All relevant approvals, documentation, or test results can be added to the DAO's ledger. Auditors or regulators can later verify these records with confidence that they have not been altered.

While this approach does not replace legal oversight, it streamlines compliance checks and fosters proactive identification of non-compliant work.

### *5.5.2 Quality Control via Sensors and IoT*

Recent trends in the built environment see increased deployment of sensors—monitoring factors such as vibration, concrete cure rates, or temperature. When integrated into a DAO via oracles:

1. **Real-Time Alerts**: If an IoT sensor detects anomalies that suggest substandard materials or structural weaknesses, an automated trigger can raise an alert within the DAO.
2. **Proof of Delivery**: Sensors can confirm the arrival and condition of materials. Combined with geolocation data, they also reduce disputes about late deliveries.
3. **Automated Proposals**: Upon detecting critical issues, the DAO might automatically freeze payments until an approved inspector validates corrective measures.

By using ML to interpret sensor data, further automation becomes possible. For instance, predictive models could pinpoint areas likely to fail safety checks, prompting pre-emptive interventions.

## 5.6 Case Studies and Practical Use Cases

### 5.6.1 *The DAO*

The DAO was an early experiment in decentralised autonomous organisations, operating as an investor-driven venture capital fund on the Ethereum blockchain. Launched in April 2016 through a token sale, it quickly gained prominence as one of the largest crowdfunding efforts in history. The DAO's premise was to allow token holders to collectively invest in both commercial and non-profit projects without relying on a traditional management team or board of directors. Its open-source code facilitated transparent governance and decision-making, all underpinned by smart contracts.

However, in June 2016, hackers exploited a weakness in the DAO's smart contract code, siphoning off roughly one-third of its Ether holdings. In a controversial move, most of the Ethereum community implemented a hard fork of the blockchain to recover the stolen funds, effectively splitting the network into two: Ethereum (which restored the funds) and Ethereum Classic (which continued with the original, unforked chain). Following the hack and the subsequent fork, the DAO's token was delisted from major exchanges by September 2016. Despite its short lifespan and eventual shutdown, the DAO's story remains a pivotal lesson in smart contract security and the governance complexities inherent in decentralised systems (The Economist, 2016).

### 5.6.2 *Selection of Use Cases*

A selection of illustrative scenarios can help construction professionals envision how DAOs and AI might operate in practice.

#### *Hybrid DAO for Procurement*

**Context**
A mid-sized construction consortium struggles with inefficient material ordering and delayed payments to suppliers. They establish a DAO to manage procurement for a single project phase (e.g., structural steel).

**Implementation**

1. **Budget Allocation**: The consortium stakes project funds in a multi-sig DAO treasury.
2. **Proposal Process**: Project managers propose material orders on-chain, specifying supplier, quantity, and cost.

3. **AI Integration**: An ML model references current and historical pricing data, recommending whether to purchase immediately or await favourable price conditions.
4. **On-Chain Voting**: Consortium members holding governance tokens vote on each proposal. Upon approval, a smart contract releases funds.

**Outcome**
Payments expedite once tasks are verified, fostering trust among suppliers. Although some staff find new technologies initially confusing, the transparency reduces billing disputes. Over time, the consortium extends the DAO to additional project phases.

#### *Quality Control and Compliance*

**Context**
A large residential development must comply with local environmental regulations. The developer, council representatives, and environmental consultants form a DAO to oversee sustainability targets—e.g., carbon emissions associated with materials.

**Implementation**

1. **IoT Sensors**: Smart metres record power usage on-site, feeding data into an oracle.
2. **DAO Governance**: The developer commits to limiting emissions, so if sensor data exceed set thresholds, funds allocated for offset measures are unlocked.
3. **ML Predictions**: An AI model forecasts monthly emissions trends, prompting the DAO to vote on operational changes (e.g., adjusting equipment usage or investing in renewable generators).

**Outcome**
The local council trusts real-time, tamper-proof data, speeding regulatory approvals. The project receives favourable press for its transparent commitment to sustainability. However, the developers must refine sensor calibration after some early false alerts.

#### *Risk Pooling DAO*

**Context**
Multiple subcontractors worry about cost overruns stemming from supply chain volatility (such as fluctuating steel prices). They unite in a DAO to pool risk across the project.

### Implementation

1. **Risk-Modelling AI**: The DAO uses ML to track commodity prices and forecast disruptions.
2. **Token-based Contribution**: Each subcontractor contributes a specified quantity of tokens to a shared treasury.
3. **Trigger Mechanism**: If the AI predicts severe price hikes, the DAO automatically drafts a proposal to disburse partial subsidies to the affected trade(s).

### Outcome

This collaborative risk strategy reduces sudden cash-flow crises. While some participants feel uncertain about trusting an algorithm to assess commodity cycles, transparent on-chain data and voting records gradually build confidence.

## 5.7 Conclusion

DAOs provide a decentralised model for governance and resource allocation that has distinct advantages for the construction industry. By reducing middlemen, automating payments, and fostering transparent decision-making, they streamline project workflows plagued by multi-stakeholder coordination challenges. Combined with AI and ML, DAOs become even more powerful: they not only share verifiable data but can autonomously act on predictive insights to mitigate risk, enforce compliance, and boost efficiency.

Despite these benefits, adoption remains complex. Legal uncertainty, the learning curve for smart contracts, and the need to carefully define AI thresholds (to avoid undue reliance on algorithmic triggers) are significant considerations. In practice, a measured, hybrid approach—deploying DAOs for specific financial or auditing tasks—often yields early wins. Once stakeholders witness tangible improvements in trust and operational speed, a broader roll-out can follow.

As construction 4.0 accelerates, merging digital twin data, AI analytics, and DAO-based governance may well redefine how major projects come together, potentially forging a more collaborative and transparent future for the built environment.

Chapter 6

# Blockchain and Building Information Modelling (BIM) Ecosystems

## 6.1 Integration of Blockchain with BIM

### *6.1.1 Rethinking "BIM as Usual"*

The AECO sector has, over the last two decades (2005–2025), made significant strides towards digitalising its processes, driven largely by BIM adoption. Nonetheless, conventional BIM workflows often rely on centralised CDEs, where project data is collected and managed by a single platform or service provider. This model can inadvertently create new data silos and trust barriers, particularly when multiple stakeholders with differing objectives, software tools, and security requirements collaborate on complex projects. As discussed in the Construction Blockchain Consortium's working paper on Distributed BIM, conventional BIM approaches – although transformative in their own right – have not yet fully resolved deeper, systemic issues related to data ownership, version control, confidentiality, and incentives for high-quality data exchange (CBC, 2021).

Blockchain offers a decentralised mechanism to address many of these issues by distributing the ownership and verification of project data. Rather than placing full trust in a single data host, participants record interactions and transactions (e.g., design updates, approvals, or payments) on a shared ledger. This ledger can enforce consensus rules that protect data integrity and provide transparency. Contrasted with the centralised 'single source of truth' model common to most CDEs, blockchain's distributed approach aims to create an environment in which that 'truth' is agreed upon – and auditable – across a network of equal peers.

### *6.1.2 Drivers for a Blockchain-BIM Ecosystem*

1. **Fragmentation and Trust Deficits**
   The AECO industry is characterised by fragmented supply chains, multiple contractual relationships, and frequent scope changes. These factors undermine trust and collaboration, an issue repeatedly highlighted in

DOI: 10.1201/9781003518303-7

industry reports dating back to the Latham Review (1994). By introducing an immutable ledger where each transaction or design update is cryptographically linked to a project participant, blockchain offers an audit trail that can deter fraudulent behaviour, reduce disputes, and enhance accountability.

2. **Tamper-Evident Data Management**
   Conventional BIM relies on versioned files, often stored in a single repository. File-based versioning is prone to human error, data duplication, or malicious tampering. Blockchain's consensus-based approach ensures that every update is time-stamped and validated by the network before it is appended to the ledger. Once recorded, it becomes extremely difficult to alter past states without alerting the entire network. This property can reinforce the trustworthiness of BIM data – both geometric models and any metadata (e.g., cost estimates, schedule updates, or design liability logs).
3. **Incentivising Better Data Quality**
   One major challenge in digital construction workflows is incentivising project participants to provide accurate, timely, and high-fidelity data. Blockchains can introduce token economies that reward 'good data' or penalise incomplete or erroneous contributions. For instance, tokens might be awarded for meeting data-drop deadlines in the design or fabrication stages, thus aligning the interests of diverse project participants. [CURRENT]CBC-WP_D-BIM
4. **Futureproofing and Ecosystem Flexibility**
   BIM is increasingly deployed across the entire asset lifecycle, extending into operation and maintenance. As Digital Twins and advanced analytics gain traction, the volume and complexity of data grows exponentially – potentially outliving even the firms that created it. Blockchain's decentralised architecture, combined with peer-to-peer file-sharing protocols, can offer a more resilient framework for the long-term preservation and retrieval of asset data, even if a CDE provider ceases operations.

### *6.1.3 Potential Blockchain Architectures*

Not every blockchain is suited to BIM integration. Broadly, we can distinguish:

- **Public, permissionless blockchains (e.g., Ethereum)**, which are open to anyone but can suffer from performance bottlenecks and high transaction costs.
- **Permissioned blockchains (e.g., Hyperledger Fabric)**, which restrict participant access and can process transactions more quickly but trade away some decentralisation benefits.
- **Hybrid or consortium models**, which combine aspects of public and private blockchains to suit the complexities of AECO workflows, including compliance with legislation on data ownership and privacy.

From a practical standpoint, permissioned or consortium-based models may align more directly with the multi-stakeholder environment of construction, as they allow fine-grained control over read/write privileges while still distributing trust.

### 6.1.4 *Overcoming Barriers*

Deploying a Blockchain-BIM ecosystem will inevitably meet both cultural and technical barriers. Stakeholders accustomed to centralised data ownership may fear relinquishing control. Similarly, the legal enforceability of Blockchain-based records remains a grey area, as some jurisdictions have not fully recognised their validity in litigation or compliance contexts. Moreover, the high energy demands of certain consensus algorithms (e.g., Proof of Work) may conflict with industry-wide sustainability goals.

Nonetheless, ongoing research – such as the Construction Blockchain Consortium's white papers – indicates that alternative consensus models (e.g., Proof-of-Stake, Proof of Authority, or DAG-based) can reduce environmental impact and scale more effectively to meet the demands of large AECO projects. The next step lies in building robust pilot projects that address these hurdles through a combination of technical solutions and business-process innovation.

## 6.2 Enhancing Project Management and Decision-Making

### 6.2.1 *The Promise of Decentralised Governance*

BIM adoption has significantly improved coordination among teams, but project governance remains heavily dependent on contracts, manual oversight, and hierarchical approvals. In a blockchain-powered BIM environment, certain aspects of project governance – such as cost approvals, scheduling milestones, or submittal checks – can be codified into 'smart contracts': self-executing rules triggered when specific conditions are met.

- **Automated Payments:** Project owners could 'lock' funds in a smart contract, disbursed automatically upon verification (by sensors, digital checks, or team sign-off) that a milestone has been achieved. This ensures that subcontractors and suppliers receive prompt payment, strengthening trust and reducing project cash-flow risks.
- **Real-Time Performance Dashboards:** Linking IoT or site sensors to Blockchain ensures that project metrics – e.g., the volume of materials delivered, quality checks passed – are continuously and transparently reported. This data can feed into AI-based dashboards, enabling informed decision-making on scheduling, resource allocation, or risk mitigation.

### 6.2.2 *Streamlining Design Coordination*

Complex BIM models often originate from multiple authoring tools used by different disciplines such as architecture, structural engineering, and MEP. The result is a federated model that can present conflicting or inaccurate data if updates are not properly tracked. Blockchain-based transaction records help coordinate design changes:

- **Immutability**: Each design submission is hashed, time-stamped, and recorded. If a party modifies the model geometry or related data, other participants can verify whether the updated version is legitimate.
- **Clear Attribution**: Designers and engineers can digitally 'sign' their changes, preventing confusion about who owns or is liable for a particular segment of the model.
- **Automated Conflict Resolution**: Smart contracts can detect clashes between disciplines—e.g., structural beams intersecting mechanical ductwork—and either flag them for review or invoke automated rule-checking routines.

### 6.2.3 *Risk Allocation and Liability Management*

In current practice, it can be challenging to ascertain how design liability is allocated among multiple consultants, each operating on different versions of a model. By tying each incremental update on the blockchain to a cryptographic signature, the sequence of design decisions and responsibilities becomes evident. Should disputes arise, the ledger serves as an auditable record of who made which changes and when, reducing the time and cost of litigation.

A corollary benefit is that errors or omissions can be swiftly traced back to their source. Although many project teams are cautious about creating unalterable evidence, the impetus towards better data quality, version control, and shared accountability can, over time, foster a more collaborative culture.

### 6.2.4 *Improved Decision Support Through Data Analytics*

Blockchain-based project data can feed into advanced analytics or AI/ML models. If each record is guaranteed to be correct and time-stamped, data-mining algorithms can extract trends or predict outcomes with greater confidence. For instance, historical data might show that certain prefabrication methods generate fewer on-site conflicts or rework. This knowledge, derived from multiple previous projects, could be embedded into the design-stage decision-making for new projects, helping reduce risks and improving productivity.

## 6.3 Forms and Certification Automation

### *6.3.1 Requests for Information (RFI) Processes: Current Shortcomings*

RFIs and their chain of response and mitigations are central to communication in construction projects. They clarify design details, address ambiguities, and record queries raised by site teams. Traditionally, an RFI entails emailing questions to design professionals or uploading them to a project management portal. Delays can arise when these questions are sent late, misrouted, or answered incompletely, leading to schedule overruns or cost escalations.

In a blockchain-enabled environment, an RFI becomes a transaction:

1. **Creation**: The site engineer logs the query, referencing specific model elements from the BIM.
2. **Broadcast**: The RFI is recorded on-chain, ensuring a permanent and tamper-evident request.
3. **Response**: Relevant parties (designers, managers) receive notifications and post replies. Any update or revision is appended as a new on-chain record.
4. **Consensus**: Once resolved, the RFI can be digitally 'closed' by a consensus mechanism—e.g., a required number of signatures from design leads—to confirm that the query has been satisfactorily addressed.

### *6.3.2 Certification and Compliance in a Common Data Environment*

Certification processes (e.g., signoffs on design changes, regulatory compliance checks) can similarly be made more robust by using blockchain transactions to track each step. For example, building inspectors, quantity surveyors, and other certifiers can each sign off within the ledger. Because the sign-off is immutable, no party can retroactively alter or delete it without being detected by the entire network.

### *6.3.3 Reducing Administrative Overheads*

Conventional RFI and certification workflows can generate vast amounts of documentation: email threads, PDF attachments, meeting minutes, etc. Blockchain-based methods consolidate all that metadata into a shared ledger accessible via a CDE interface. The overhead involved in version control, manual approvals, or verifying who received which document is thus reduced. Auditing for compliance becomes more efficient, since all relevant actions and signatories are recorded in chronological, immutable sequences.

### 6.3.4 *Case Examples of Automated RFI Resolution*

While widespread adoption is still emerging, early prototypes in research and practice show how automated RFI resolutions can be built on top of a permissioned ledger. For instance:

- **Trigger-Based Alerts**: A site engineer logs an RFI referencing a specific BIM element. The Blockchain ledger triggers notifications to design consultants, prompting them to respond within an agreed timeframe.
- **Automatic Escalation**: If a query remains unresolved, a smart contract can escalate it to senior management, ensuring that no item "slips through the cracks."
- **Consensus Closure**: Once the required design lead and the project manager have confirmed that the RFI is adequately answered, the query is "closed" on-chain—documenting the final consensus and eliminating uncertainties about incomplete signoffs.

## 6.4 IP Management

BIM artefacts are valuable IP. Authors invest time and expertise to create accurate 3D models, parametric objects, or data-rich library components. In traditional, centralised environments, IP can be copied or modified without proper attribution. Over the life of a project, ownership boundaries may be blurred if multiple contractors share partial updates to the same model file.

Blockchain-based BIM approaches offer a solution:

1. **Immutable Attribution**
   Each contributor can append digital signatures and ownership tags to newly created or edited model elements. If a subcontractor or designer modifies that element, the ledger automatically logs the updated "chain of custody."
2. **Smart Contracts for Licensing**
   In some cases, design firms may grant usage licences for proprietary objects—such as prefabricated components or specialist parametric modules—on condition of fees or specific usage rights. A smart contract can automate such licensing terms. For example, once a usage fee is received, the ledger grants decryption keys or read/write permissions for certain BIM objects.
3. **Protection from Unauthorised Replication**
   While no technical method can wholly prevent data from being copied, Blockchain-based systems can prove originality and date-stamp content. If a copy or altered version emerges in a dispute, on-chain records can verify who possessed the original design at each timestamp.

4. **Incentivised Collaboration**
   When each party's IP is respected and traceable, project teams are likelier to share high-quality data. This open exchange can be further encouraged through a token-based rewards system, which acknowledges the real value contributed by each participant.

## 6.5 Data Governance and Management

Effective data governance in multi-CDE project environments underpinned by blockchain requires robust frameworks that harmonise data integrity, access control, interoperability, and compliance. Drawing from current research, below we outline key governance strategies and system architectures.

1. **CDE with Blockchain as the Trust Layer**
   To support multiple platforms such as Autodesk Construction Cloud and Bentley ProjectWise, blockchain can provide a decentralised registry of unique asset identifiers and metadata. This enables transparent cross-platform references and traceability without centralisation. OpenCDE frameworks note that this federated layer preserves platform autonomy while facilitating secure data exchange (Werbrouck et al., 2023).
2. **Standards and Interoperability**
   Combining BIM with blockchain demands adherence to standards like IFC, ISO 19650, and common metadata schemas to ensure semantic consistency across environments. Reviews underline that formal data schemas and shared object identifiers are prerequisites for automated validation across CDEs (Xu et al., 2024).
3. **Access Rights and Permissioning**
   Permissioned blockchains (e.g., Hyperledger Fabric) can support role-based controls, enabling each stakeholder to manage permissions granularly. This aligns with GDPR compliance and data-privacy best practices by restricting ledger access and usage to authorised roles (Kloosterman et al., 2024).
4. **Immutable Accountability and Audit**
   Every change or RFI, model update, or inspection sign-off can be time-stamped and linked to a digital signature. High-impact contributors such as Zheng et al. (2019) demonstrate how this structure improves accountability, ensuring each stakeholder can be verified and audit trails are undeniable (Xu et al., 2024).
5. **Scalability and Off-Chain Data Handling**
   Given large BIM datasets, solutions combining on-chain metadata with off-chain storage (e.g., IPFS or cloud) provide practical scalability. Semantic Differential Transaction techniques store only model deltas on

the ledger—reducing storage overhead while maintaining provenance and enabling full reconstruction (Werbrouck et al., 2023).

6. **Organisational and Legal Integration**
   Transitioning from siloed CDEs to a federated, Blockchain-based governance model will call for:

   - **Standardised data taxonomy** for consistent cross-CDE referencing
   - **Updated contract clauses** covering ledger custody, IP rights, access permissions, and compliance
   - **API and integration protocols** to enable secure read/write interactions with the Blockchain layer

   These considerations mirror recommendations from empirical BIM–Blockchain research emphasising stakeholder coordination and legal alignment

By combining federated identity, standardised metadata, permissioned ledgers, immutable audit trails, efficient off-chain strategies, and contractual governance, Blockchain-based data architecture enhances interoperability across CDEs. This supports secure, auditable, and scalable BIM ecosystems capable of meeting the requirements of Construction Digital Transformation.

## 6.6 Higher Complexity CDEs

Large construction projects commonly rely on a host of digital collaboration platforms like Bentley ProjectWise, Autodesk Construction Cloud and ASite, holding design and delivery data with often many other specialised portals. While each system promises a "single source of truth," in practice, multiple partial truths arise when different teams store data in separate proprietary environments. Data duplication, mismatched file versions, and complicated user permissions create friction that undermines the promise of BIM interoperability.

**Blockchain as a Federating Layer**
Rather than forcing all teams onto a single CDE, blockchain can operate as a neutral "federating layer" that synchronises key project data and events across different platforms:

**Cross-Platform Identification**
Assets, objects, or issues gain unique IDs stored on the ledger, ensuring that the same item referenced in ProjectWise or Autodesk is clearly matched. This promotes consistency, even if each CDE uses a different internal naming structure.

**Granular Access Control**
Permissioned Blockchains allow each CDE environment to dictate precisely which data is exposed (e.g., read-only for some participants, read/write for

others) and which remains private. This fosters a robust system of data governance that can track usage rights or licensing conditions across multiple platforms.

**Standard Methods and Procedures**
By embedding standard BIM procedures (e.g., ISO 19650 principles) directly into the ledger's validation rules, each new file version or data transfer must meet the project's agreed quality control checks. If it fails these checks, the blockchain network rejects the transaction or flags it for review.

**Compliance and Auditing**
Because the ledger timestamps changes, managers can easily audit who made updates, when they were made, and why. This is particularly beneficial for verifying compliance with contract protocols or regulatory mandates. Where applicable, external auditors or regulatory bodies can be granted permissioned access to track compliance in real time.

**Organisational and Contractual Considerations**
Implementing a blockchain federation of multiple CDEs also entails organisational change. Project participants will need to:

**Agree on Data Standards**: For example, define a uniform classification system or naming convention so that cross-referencing remains accurate across platforms.
**Revisit Contract Clauses**: Adjust T&Cs around data ownership, usage rights, liability for on-chain errors, and responsibilities for maintaining the ledger.
**Adopt Interoperable APIs**: Integrate each CDE with the Blockchain network via robust application programming interfaces that can read and write ledger transactions seamlessly.
**Uphold Cybersecurity**: Permissioned networks still require strong governance—assigning user roles, rotating cryptographic keys, and employing zero-trust security approaches.

## 6.7 Use Cases and Pilots

Though blockchain in construction remains an emerging domain, several pilot projects and academic–industry collaborations are beginning to illustrate the technology's potential. Summaries of notable examples include:

1. **Automated Payment and Supply Chain Assurance**
   A UK-based infrastructure project tested a pilot in which subcontractors received milestone payments via smart contracts, reducing average waiting times by 35%. Materials were tracked via QR codes that linked

each delivery record to the ledger, enabling real-time verification of their provenance.

2. **Federated Model Coordination**
   A consortium-based project with multiple engineering firms used a permissioned ledger to unify data from Autodesk Construction Cloud and Bentley ProjectWise. Each design change triggered a ledger transaction capturing the new revision, which drastically reduced confusion about which model version was authoritative.
3. **Regulatory Inspections and Approvals**
   Municipal authorities in a European pilot used a private blockchain network to conduct building code reviews. Inspectors' signoffs were added to the ledger as cryptographically signed transactions. Projects reported shorter turnaround times for compliance approvals and fewer disputes over "who signed off on what and when."
4. **Digital Twin and Asset Management**
   An owner-operator trialled integrating blockchain with an in-house digital twin platform, feeding sensor data about energy usage into a distributed ledger. Smart contracts automatically notified facility managers if consumption exceeded thresholds, rewarding building users who adhered to sustainability guidelines. Over time, this real-time data created a robust performance history, key to refining future designs.
5. **IP Tokenisation and Collaboration**
   A design–build contractor developed parametric "BIM objects" for offsite fabrication. By encoding these objects as non-fungible tokens, the contractor could licence or share them with sub-parties while safeguarding ownership. Should a sub-party incorporate the object into a new BIM design, the ledger automatically attributed and time-stamped the usage.

These examples, though relatively small in scale or scope, demonstrate how blockchain can address recurring industry pain points such as payment delays, version control errors, and data security gaps.

## 6.8 Conclusion

Blockchain's integration with BIM has the potential to transform the AECO sector by mitigating the perennial challenges of trust, interoperability, and fragmented workflows, as discussed:

- **Governance and Automation**: Smart contracts can streamline approvals, accelerate payments, and codify collaborative rules, reducing administrative overheads and project disputes.
- **Data Integrity**: Immutable, tamper-evident logs confirm exactly when and by whom changes were made, greatly simplifying compliance checks and liability allocation.

- **Distributed Collaboration**: Rather than forcing all parties onto a single CDE, Blockchain can be a federating layer, harmonising data and processes among multiple platforms—such as Bentley ProjectWise and Autodesk Construction Cloud—without sacrificing existing systems.
- **IP**: Immutable, traceable records and token-based licensing schemes encourage higher-quality data-sharing while protecting the authors' rights to their BIM artefacts.
- **Outlook**: Larger-scale adoption will require robust standards, validated legal frameworks, and pilot projects that demonstrate value across the construction supply chain. Emerging consensus models (e.g., Proof-of-Stake, DAG-based approaches) and evolving data-privacy solutions (e.g., zero-knowledge proofs) will likely help resolve current concerns over sustainability, data privacy, and performance.

Through a considered approach, embracing both technical deployments and new business processes, blockchain-infused BIM stands to underpin the next wave of Construction 4.0 innovation—one that not only enhances current workflows but also fosters a more integrated, reliable, and sustainable built environment for the decades to come. By building on insights from the Construction Blockchain Consortium and related initiatives, industry leaders can seize the opportunity to revolutionise how projects are delivered and managed, aligning strategic, technical, and contractual frameworks in a more transparent and collaborative ecosystem.

Chapter 7

# Blockchain in Digital Twins for Carbon Targets

## 7.1 Background

A *digital twin* is a virtual model or simulation of a real-world physical object, process, or system that mirrors its state and behaviour based on real-time data (IBM, 2021; LEAP:IN, 2022). The concept, originally popularised in the manufacturing and aerospace sectors, has broadened to include assets like buildings, power grids, supply chains, and even entire cities. A digital twin continuously ingests data from sensors and other sources on the physical "twin," enabling up-to-date reflections of its condition and performance. For example, a wind turbine's digital twin would receive live data on energy output, temperature, vibration, etc., allowing operators to simulate scenarios and predict maintenance needs (IBM, 2021).

Unlike static simulations, digital twins maintain a two-way connection: they not only *receive* data from the physical asset but can also send back insights or control inputs, creating a feedback loop for optimisation (IBM, 2021). This real-time synchronisation and the ability to run simulations make digital twins powerful for informed decision-making. In summary, the digital twin serves as a *living digital counterpart* of a physical entity, enabling users to monitor and test changes in a virtual environment before applying them.

### *7.1.1 Carbon Targets and Their Importance*

*Carbon targets* refer to the specific goals set by governments, industries, or organisations to reduce greenhouse gas emissions (primarily carbon dioxide) in line with climate change mitigation efforts. These targets can be absolute emission reduction goals (e.g. reduce 50% of emissions by 2030 relative to a baseline) or intensity targets (e.g. emissions per unit of GDP or product). A prominent form of carbon target is the commitment to achieve "net-zero" emissions by a certain date (meaning any remaining emissions are balanced by removals or offsets).

The importance of such targets is underscored by climate science: to limit global warming to 1.5°C above pre-industrial levels – the aspirational goal

DOI: 10.1201/9781003518303-8

of the 2015 Paris Agreement – global $CO_2$ emissions need to reach net-zero around mid-century (Cozzi and Gould, 2019). Exceeding 1.5–2°C of warming dramatically increases the risks of severe climate impacts (extreme weather, sea level rise, ecosystem loss) (Cozzi and Gould, 2019). Thus, nearly all major economies have adopted carbon targets; by late 2019, at least 65 jurisdictions (including the EU and UK) were actively considering or had set long-term net-zero goals for 2050 or sooner (Cozzi and Gould, 2019). These targets serve as critical benchmarks to drive policy and innovation. They inform international agreements (like the Paris Agreement commitments), national legislation (for instance, the UK's legally binding net-zero 2050 target), and corporate sustainability strategies. Carbon targets also signal to markets and the public that concerted efforts will be made to transition away from high-carbon processes (such as fossil fuel combustion) towards low-carbon and renewable alternatives. In essence, carbon targets are important because they provide a *quantitative framework* for tracking progress in combating climate change and because they hold actors accountable to making measurable emissions reductions.

### 7.1.2 Carbon Management with Blockchain and Digital Twins

Managing carbon emissions to meet these targets presents significant data and governance challenges. Organisations must accurately measure emissions across complex systems (factories, supply chains, cities) and report them in a transparent and verifiable way. Here, the integration of blockchain with digital twins emerges as a promising solution. A digital twin can simulate and monitor the energy use and emissions of a system in real time, but ensuring the *integrity* of the twin's data and sharing it with stakeholders requires trust.

Blockchain can provide that trust layer. By recording key pieces of the digital twin's data (such as carbon output readings, energy consumption, or process optimisations) onto an immutable ledger, stakeholders can be confident that the data has not been manipulated or falsified. This is especially significant for carbon management, where there may be incentives to under-report emissions. For example, integrating blockchain can mitigate risks of *misrepresentation* of emissions data by creating a tamper-evident audit trail. It "establishes a transparent and immutable ledger that holds all participants accountable, ensuring that the data presented is accurate and trustworthy" (Fitriawijaya and Taysheng, 2024). Moreover, blockchain's smart contracts (self-executing code on the ledger) can automate certain aspects of carbon management, such as issuing alerts or even tokenised carbon credits when emissions fall below or exceed thresholds.

In sum, combining digital twins with blockchain enables a secure, transparent monitoring system for carbon performance. The digital twin provides the *data* and simulation capability, while blockchain provides *data governance* – guaranteeing integrity, enabling data sharing without a central authority, and facilitating automated verification. This synergy is significant because it can bolster confidence in reported progress towards carbon targets among regulators, investors, and the public. It turns what might otherwise be internal data into a *trusted record* of carbon footprint, thus marrying the operational benefits of digital twins with the transparency and trust of blockchain. The subsequent sections delve deeper into how this integration works and its real-world applications for tracking and meeting carbon targets.

## 7.2 How Blockchain Enhances Digital Twins

Blockchain technology can enhance digital twin systems in several key dimensions: **data integrity and security, transparency, automation via smart contracts**, and **decentralised trust**. Each of these aspects addresses common challenges in managing and sharing digital twin data, especially in multi-stakeholder environments like supply chains or smart cities.

### 7.2.1 *Data Integrity, Security, and Transparency*

Digital twins rely on continuous streams of data (from IoT sensors, databases, user inputs) to update the virtual model. Ensuring this data is reliable and untampered is crucial – otherwise the twin's insights or simulations could be invalid. Blockchain's core functionality as an immutable ledger directly strengthens data integrity. Once digital twin data (such as an emissions reading or a maintenance record) is recorded on a blockchain, it cannot be altered retroactively without consensus from the network, which is typically infeasible. This immutability means historical data feeding the twin is *frozen in time* and auditable. For example, a factory's energy consumption logs or carbon emission measurements could be periodically hashed and stored on blockchain; later, auditors can verify those records against the twin's outputs to ensure consistency.

In the context of carbon tracking, researchers note that blockchain provides a *transparent and auditable verification process* for emissions data, building trust among stakeholders and regulators. Security is also enhanced: blockchain hashing and digital signatures secure data entries, also authenticating the digital twin models and algorithms themselves – stakeholders can confirm that the twin's computational model has not been tampered with by checking its hash or signature on the ledger. This is important over the twin's lifecycle to ensure consistency of the model. With data distributed across many nodes, there is no single point of failure for cyberattacks, improving resilience. Equally important is *transparency*.

In a blockchain, all authorised participants (or the public, in an open blockchain) can view the relevant transactions/data entries (Saxena, 2024). This means that the data underlying a digital twin's conclusions is open to examination, which is especially useful when multiple parties contribute data to or rely on the twin. For instance, in a shared infrastructure project, an IoT sensor's readings logged via blockchain are visible to the asset owner, operators, regulators, etc., reducing information asymmetry. According to the European Commission, blockchain is a powerful tool to improve the transparency, accountability and traceability of greenhouse gas emissions data, yielding more accurate and reliable information on carbon footprints (European Commission, 2024). Thus, by using blockchain, digital twins become part of a *transparent data ecosystem* where integrity is guaranteed, and security is built in.

### 7.2.2 *Smart Contracts and Automation*

One of the most impactful features of modern blockchains (particularly those like Ethereum or Hyperledger Fabric) is the ability to deploy smart contracts – self-executing code that runs when predetermined conditions are met. Integrating smart contracts with digital twins enables automation of responses or workflows based on the twin's data. For example, consider a digital twin of a building that tracks energy usage and emissions in real time. A smart contract could be set up to automatically trigger an *alert or action* if the building's carbon emissions exceed a set target for the month. When the twin's data (energy metres, HVAC usage, etc.) indicates an emissions threshold is passed, it could send a transaction to the blockchain, activating a smart contract that perhaps notifies engineers or even purchases carbon offsets from a marketplace.

In an industrial context, researchers have proposed architectures where digital twins feed data to smart contracts which then handle decision logic and record events. One such framework for structural health monitoring integrates sensors, digital twins, and smart contracts so that if a sensor detects an anomaly, a blockchain record is created and a smart contract automatically notifies stakeholders and recommends maintenance actions. The smart contracts essentially encode business rules or sustainability policies. This automation reduces the need for manual data processing or oversight, making the system more efficient. It also adds *consistency* – the rules execute the same way every time conditions are met, without relying on human intervention. For carbon management, smart contracts could automate carbon accounting (e.g., tally emissions from various sources recorded on-chain and compare against a target) or automate carbon credit transactions. For instance, a company might use a smart contract to automatically retire a tokenised carbon credit when its digital twin indicates one ton of $CO_2$ has been emitted beyond a cap. Indeed, blockchain-based **decentralised applications (DApps)** can be built on top of digital twin data to handle complex logic. These DApps use the twin as an information source and the blockchain as the execution and record-keeping environment. Industry examples are emerging: Kongsberg

Digital notes that *accurate data from the digital twin can be delivered into smart contract networks with blockchain technology for auditable results across the value chain* (Hansen, 2022)– illustrating how smart contracts can consume twin data and automatically produce audit logs or compliance reports.

In summary, smart contracts turn the digital twin + blockchain combination into an *active system* – not only reflecting the world, but also initiating predefined actions (from notifications to financial transactions) in response to the twin's state. This level of automation is highly valuable for enforcing carbon targets and sustainability rules without constant human monitoring.

### 7.2.3 Decentralisation and Trust-Building

In many scenarios, the full benefit of digital twins (especially on a city or national scale) comes when data is shared across organisations. However, data sharing is often impeded by lack of trust – organisations might be reluctant to give a central entity control over their data or might worry that data could be altered or misused. Blockchain introduces a decentralised trust model that can overcome these barriers. Because no single party controls the ledger, all participants are on a more equal footing and can trust the system rather than having to explicitly trust each other. This is crucial when multiple stakeholders (e.g., different companies in a supply chain, or various departments in a government) need to collaborate on a unified digital twin of a complex system like a supply network or a city infrastructure. The *decentralised consensus* mechanism ensures that updates to the data are agreed upon by the network, preventing any one participant from unduly influencing the shared data.

As the UNFCCC's Climate Change secretariat has noted, blockchain's decentralised nature can *"foster greater transparency...and help bring trust... in the fight against climate change"*(UNFCCC, 2021a). In practice, this means a digital twin that spans organisational boundaries (for instance, a national digital twin of an energy grid with many energy providers) can use blockchain so that each provider writes its data to a common ledger. All providers and regulators can trust this shared ledger as the definitive source, rather than maintaining siloed databases and reconciling data via audits. Decentralisation also improves *collaboration*: since the blockchain network can be permissioned to include all relevant stakeholders, it creates a collaborative platform for data exchange. The UK's National Digital Twin initiative envisions an "ecosystem of connected digital twins" with secure data sharing between infrastructure operators (IET and Atkins, 2019). Blockchain is seen as an enabler here, by providing the security and federation needed for that data exchange.

Furthermore, blockchain can incorporate **self-executing governance** – using smart contracts for access control and data-sharing rules – which reduces the need for a centralised administrator of the digital twin data. All these factors help build trust in the data and in the overall system. When participants are confident that data is accurate (integrity), visible (transparency), and that the rules are fair and automatic (smart contracts), they are more likely to participate in data-sharing ecosystems. In the context of carbon targets, this trust means companies or units are more willing to contribute their emissions data to a shared platform (like an industry-wide digital twin for carbon footprint) because they know the data cannot be manipulated and their contributions are recorded fairly.

In summary, blockchain's decentralisation addresses the *trust gap* in multi-stakeholder digital twin deployments, turning the twin from a single-entity tool into a *collaborative network asset*. This trust-building aspect is perhaps the most pivotal enhancement, as it unlocks large-scale use of digital twins for climate action beyond the confines of one company or agency.

## 7.3 Application of Blockchain in Carbon Target Tracking

By combining blockchain with digital twins, innovative applications are emerging to monitor, verify, and incentivise reductions in carbon footprints. Key use cases include reliable tracking of emissions data, tokenisation of carbon assets (like credits or allowances), and sector-specific solutions in energy, transportation, and manufacturing.

- **Monitoring and Verifying Carbon Footprints:** A fundamental step towards meeting any carbon target is accurately monitoring emissions and verifying those measurements. Digital twins can model complex processes (a factory's operations, a building's energy usage, a city's traffic flow) and estimate associated carbon emissions in real time. By integrating blockchain, each data point in the carbon footprint calculation can be *logged and verified*. Blockchain serves as a **tamper-proof carbon ledger**. For example, in an industrial facility, IoT sensors might feed the digital twin data on fuel consumption or output levels, which the twin uses to calculate $CO_2$ emitted. Important milestones or aggregate emissions data can be recorded on blockchain. Because these records are immutable and timestamped, they provide a trustworthy audit trail for corporate sustainability reports or regulatory compliance. If an auditor or regulator wants to verify the facility's reported emissions, they can compare the report against the blockchain log. This greatly reduces the opportunity for *greenwashing* or accidental errors. An illustrative real-world application is by Hyundai Motor Company and Kia Corporation, which in 2023 introduced a blockchain-based carbon emissions monitoring system for their supply chain . Their Supplier $CO_2$ Emission Monitoring System (SCEMS) computes emissions at every stage of a supplier's operations and uses blockchain to assure the data's transparency and integrity (HYUNDAI, 2023). By leveraging blockchain in this way, Hyundai and Kia are able to securely and promptly collect carbon data from various partners and ensure it remains unaltered, thus confidently tracking progress towards their supply chain carbon targets. Likewise, in the building sector, researchers have integrated BIM/digital twin models with blockchain to secure the process of carbon disclosure in green buildings – ensuring that from data collection to reporting, all emission data entries are traceable and verifiable (Fitriawijaya and Taysheng, 2024). Blockchain essentially acts as the *verifier* for the digital twin's output, giving stakeholders (whether corporate executives, regulators, or the public) confidence that the carbon footprint numbers are real. Moreover, blockchain records can be made accessible to all stakeholders simultaneously, improving transparency of carbon performance. This is valuable for supply chains where each supplier's emissions contribute to the final product's footprint; a shared ledger allows all companies in the chain to see verified emissions data in real time. In summary, blockchain-enhanced monitoring turns the digital twin from an internal tool into a **platform for accountability**, where carbon footprints are not just calculated but also independently verifiable and trusted by third parties.

- **Tokenisation and Carbon Credits:** Beyond measurement, blockchain enables *tokenisation* – the representation of assets or units of value as digital tokens on the ledger. In the context of carbon management, one of the most important assets is a carbon credit (or allowance). Carbon credits, such as those from cap-and-trade systems or voluntary offset projects, represent a certified unit of emissions reduction (often one ton of $CO_2$). Blockchain can tokenise these credits, meaning each credit is a unique digital token recorded on-chain. This has several advantages for carbon target tracking and trading. Firstly, tokenisation can prevent double-counting and fraud in carbon markets. Each credit token has a unique identifier and ownership record on blockchain, so the same credit cannot be claimed by two entities at once (UNFCCC, 2021a).

  The UNFCCC reports an example where Energy-Blockchain Lab and IBM created a blockchain platform in China to trade carbon assets, allowing companies to reliably track their carbon footprints and meet quotas by buying credits from others on a shared ledger. In such a system, when a company's digital twin shows lower emissions than its target, the smart contract could automatically issue a token reward (analogous to a credit) that the company can sell; conversely, if it exceeds a cap, it might be required to purchase tokens. Blockchain's *transparency* ensures these transactions are visible and traceable, building confidence in the carbon market's integrity. Secondly, tokenisation enables automation: smart contracts can handle the retirement of carbon credit tokens when used, or enforce that tokens are only issued when verifiable emission reductions occur (for instance, linking issuance to a digital twin of a forest or renewable energy project that demonstrates carbon sequestration or avoidance). There are already use cases where blockchain tokens represent carbon offsets – for example, projects that tie tokens to reforestation efforts or renewable energy certificates. By integrating this with digital twins, the actual performance of a project can dictate token issuance. Think of a solar farm digital twin connected to blockchain: every MWh of clean energy generated (tracked by the twin) could automatically issue a token representing a certain emissions avoided, which then can be sold as an offset credit. This creates a direct, verifiable link between real-world carbon reductions and digital certificates. Additionally, tokenisation can support **carbon credit trading platforms** that operate globally and continuously. Traditional carbon accounting is periodic and often siloed; in contrast, a blockchain-based system could allow near real-time accounting and exchange. For example, if a manufacturing plant's twin shows it beat its monthly carbon target, it could instantly list surplus credits on a blockchain marketplace accessible by other firms looking to buy credits to cover their shortfall. Overall, blockchain's tokenisation capability introduces liquidity, standardisation, and trust to carbon credits, which are key for scaling carbon markets as tools to achieve targets. It also encourages *incentive mechanisms*: companies see tangible rewards (token credits) for going beyond their targets, and clear costs (needing tokens) if they fall short – all enforced by code.

- **Use Cases in Energy, Transportation, and Manufacturing:** The integration of blockchain and digital twins for carbon tracking is being explored across various sectors, each with its own specific applications:
  - **Energy Sector:** Energy systems are central to carbon emissions, and digital twins are increasingly used to manage energy infrastructure (like power grids, wind farms, buildings' energy management). Blockchain can help certify and trace renewable energy generation and usage, contributing to carbon accounting. A notable use case is peer-to-peer energy trading. With a prosumer energy digital twin (households with solar panels, batteries, EVs), blockchain can enable trading of surplus renewable energy between neighbours, with automated carbon calculations. The European Commission highlights that blockchain can track and report carbon footprint reduction across the entire energy value chain, providing instant authentication and clear records(European Commission, 2024). For instance, projects have used blockchain tokens to represent kilowatt-hours of green energy; when someone buys those tokens, they effectively buy the *environmental attribute* of that energy. A digital twin of the grid can feed data on energy flows to blockchain, ensuring tokens correspond to actual delivery of clean energy. Another case: smart grid management – as grids incorporate more distributed energy resources (solar panels,

wind, etc.), digital twins model grid behaviour, and blockchain can register device-level data to build trust in a decentralised grid control system (Almeida and Vrees, 2020). By having devices (solar inverters, smart thermostats) log their status and actions on blockchain, grid operators can trust the coordination among millions of devices needed to balance supply and demand, which indirectly helps reduce reliance on fossil peaker plants (thus cutting carbon). In summary, energy sector implementations focus on using blockchain to ensure that data about energy generation, exchange, and consumption – which digital twins use for optimisation – is transparent and credited properly in terms of carbon impact. This ranges from verifying that a data centre is running on green power to creating new marketplaces for renewable energy supported by digital twins of energy assets.

- **Transportation and Supply Chains:** Transportation (from personal vehicles to freight logistics) is another major source of emissions, including the harder-to-track *Scope 3* emissions in supply chains. Digital twins of vehicles, fleets, or logistics networks can simulate fuel use and optimise routes to cut emissions. Blockchain comes into play to provide traceability of emissions along complex supply chains. For example, consider a supply chain digital twin that models the journey of a product from raw material to store shelf, accumulating the carbon emitted at each step (mining, manufacturing, transport). If each supplier in the chain records their emissions data on a shared blockchain, the final product's carbon footprint can be transparently verified (European Commission, 2024). Hyundai and Kia's SCEMS mentioned earlier is a concrete application in automotive manufacturing supply chains – it uses AI and blockchain to accurately calculate suppliers' emissions and predict future emissions, ensuring the carmakers can identify carbon-intensive steps and work with those suppliers on improvements (HYUNDAI, 2023). Similarly, the agrifood industry is exploring blockchain for *carbon traceability*, proposing using blockchain to trace carbon footprints in agriculture supply chains (e.g., palm oil production), recording emissions at each stage from farm to factory (TraceX, 2022). Transportation systems at city scale can also benefit. A city government might have a digital twin of its transportation network (buses, traffic, trains) to reduce congestion and vehicle idling. If ride-sharing services, freight companies, and public transit all log data (like fuel consumed, passenger miles) on a blockchain linked to that twin, the city can create a unified, tamper-proof record of transport emissions, identify hot spots, and credit improvements (like increased electric vehicle usage or better traffic flow) in an open way. Another interesting use case is *green logistics*: companies like DHL or UPS could employ blockchain and IoT to track a package's carbon footprint. Each segment (air, sea, road) could add to an on-chain record via its digital logistics twin, yielding a final emission tally per package that is trusted. This data can then feed into carbon offsetting programs or customer-facing carbon labels. In short, blockchain enables **end-to-end carbon visibility** in transportation and supply chain scenarios, complementing the optimisation that digital twins provide with an *accountability mechanism*.

### *7.3.1 Manufacturing and Industry*

Manufacturing combines energy use, supply chain inputs, and processes – all sources of emissions. Digital twins are increasingly used on factory floors and in product design to improve efficiency and reduce waste. Blockchain can document these efficiency gains and waste reductions as part of carbon accounting. A notable example is BASF's reciChain project in the plastics manufacturing sector. BASF created reciChain as a blockchain-based platform using digital twin concepts to track plastics through their life cycle and incentivise recycling (BASF, 2023; Mullican, 2022).

In the British Columbia pilot, digital "badges" were attached to plastic products (a kind of digital twin identifier), and as materials moved through recycling processes, the blockchain recorded their journey, awarding tokens for successful recycling loops. This track-and-trace system with blockchain not only helps in waste reduction (circular economy) but also in calculating the carbon savings from recycling plastic versus producing new plastic. In general manufacturing, blockchain can record the provenance of materials and parts, which ties into carbon footprint when manufacturers are required to report the embedded carbon in their products.

A digital twin of a product can aggregate all the material data and processing data; blockchain can then store hashes or proofs of that data so an end customer or regulator can verify claims like "50% of this product's content is from recycled low-carbon material". Another area is **emissions compliance** in heavy industry (steel, cement). A digital twin might simulate a plant's processes to find lower-carbon operating modes. When those are implemented, blockchain can log key parameters (like kiln temperatures, fuel type used, output quality) to certify that a batch of product was made under low-carbon conditions. This could support emerging concepts like green steel or green cement certification. We also see interest in using blockchain to manage carbon trading within industrial parks or between facilities. For example, if one factory's twin shows it is under its emissions quota and another is over, a blockchain-based system could automatically facilitate a carbon credit trade between them (UNFCCC, 2021a).

The trusted data from each factory's twin (say, emissions per hour) would be the basis, and smart contracts would handle the transaction once conditions are met. Manufacturing companies are typically cautious about data sharing, but blockchain's security and permission controls can allow them to share necessary data (emissions, energy use) without exposing sensitive proprietary information. This fosters a collaborative approach to emissions reduction across an industry supply chain. In summary, manufacturing use cases revolve around trusted data exchange for carbon efficiency – whether it's tracking materials for recycling, sharing emissions data with clients, or enabling internal carbon markets, blockchain ensures the integrity of the information coming from digital twin models and sensors on the factory floor.

Across all these sectors, a common thread is that blockchain provides a *trusted record and transaction layer* on top of the digital twin's analytical layer. This combination not only helps monitor and verify current carbon performance but can also enable new mechanisms (like token rewards or automated carbon trading) that actively drive behaviour towards meeting carbon targets.

## 7.4 Case Studies and Real-World Implementations

Several companies and projects have begun implementing blockchain-integrated digital twin solutions for carbon tracking and sustainability. These case studies illustrate the practicality of the approach and offer key takeaways on benefits and challenges:

- **Kongsberg Digital's "Kognitwin" for ESG in Energy:** Kongsberg Digital (a Norway-based industrial software company) has applied digital twins and blockchain in the context of energy asset management and ESG (Environmental, Social, and Governance) reporting. In 2022, Kongsberg outlined how its *Kognitwin* digital twin platform can be combined with smart contracts on a blockchain to automate carbon emissions tracking and reporting for industrial facilities (Hansen, 2022). In a typical use case, a Kognitwin model of an offshore oil & gas facility is deployed to monitor operations. The twin provides a baseline calculation of the facility's carbon emissions and then continuously tracks deviations in real time. The key innovation is that *"accurate data from the digital twin...is available at any time, ready to be delivered into smart contract networks with blockchain technology for auditable results across your value chain"*. In practice, as the twin updates the emissions data, it triggers smart contracts that log these data on blockchain and even handle parts of the disclosure process. The immutable record ensures that the company's sustainability reports are backed by auditable data trails, reducing manual paperwork and increasing trust with regulators and investors.

  By integrating blockchain, Kongsberg's solution significantly improves the efficiency and credibility of ESG disclosure. It addresses the common pain point of ESG data being siloed and manually aggregated (often with errors). The case shows that even in heavy industries, digital transformation can enable near-real-time carbon accounting. A takeaway is that partnering operational technology (digital twin) with IT (blockchain and smart contracts) can break down data silos – Kongsberg noted that manual ESG processes led to data discrepancies and low trust, which their integrated approach overcame (Hansen, 2022). This demonstrates how transparency can be enhanced in traditionally opaque reporting processes. However, the case also highlights the need for common standards: Kongsberg pointed out that ESG lacks universal data standards and that automation wasn't leveraged enough historically, implying industry-wide collaboration is needed to realise full benefits.

- **Hyundai & Kia's Blockchain-Based Carbon Supply Chain System:** In the automotive sector, Hyundai Motor and Kia in 2023 launched a Supplier $CO_2$ Emission Monitoring System (SCEMS) that uses blockchain to manage carbon emissions data from their suppliers. This system is essentially a *decentralised digital twin of the supply chain's carbon profile*. It uses AI to estimate emissions at each stage and blockchain to secure the data. Hyundai and Kia tested the approach and verified that a high-performance blockchain could promptly and securely collect emissions data from various partners. The blockchain assures data transparency and integrity, meaning the carmakers and their suppliers share a single, trusted view of emissions.

  This case demonstrates blockchain's value in multi-actor environments – the automotive supply chain involves hundreds of companies (parts makers, raw materials suppliers, logistics providers). SCEMS shows that blockchain can serve as a neutral platform where each supplier's emissions are recorded in a trusted way, enabling the OEM (original equipment manufacturer) to accurately aggregate total supply chain emissions. One outcome is reduced reporting burden on suppliers: the system *"relieve(s) suppliers from time- and cost-consuming work"* of data reporting while enabling more accurate carbon management. This suggests a win-win: suppliers get an easier method to report and possibly benchmark themselves, while the OEM gets high-quality data. A challenge observed is ensuring all partners adopt the system and share truthful data – which is exactly what the blockchain's trust mechanism is meant to encourage. The case underlines the importance of performance and scalability: Hyundai specifically noted using a high-performance blockchain to handle data from many sources without delay. This echoes a common concern that blockchain systems must scale to enterprise transaction volumes; their successful pilot indicates that modern blockchain platforms (likely permissioned ones) can meet such demands. This implementation by major global manufacturers is a strong validation of blockchain's role in scope 3 (supply chain) emissions tracking. It hints that similar approaches could be adopted in other industries where supply chain emissions dominate (e.g., electronics, retail).
- **BASF's reciChain – Circular Economy and Carbon Benefits:** BASF, a large chemical company, developed the reciChain program, first piloted in 2020 in Canada, focusing on plastics recycling. While the primary goal is

improving recycling rates (by tracking plastic from production to reuse), the system also has carbon implications, as recycling typically reduces life-cycle emissions. In reciChain, blockchain is used to *"support the track & trace, and monetisation of plastics within the value chain"* via digital tokens and a digital twin-like tracking of plastic batches. Each piece of plastic is tagged (digitally) and as it moves through recycling, information is added to the blockchain, creating a transparent log of its journey and recycled content. Participants (manufacturers, recyclers, retailers) are part of the blockchain network.

The pilot demonstrated that they could track plastic from "pellet to pellet" – meaning from raw plastic, through product manufacturing, consumer use, and back to recycled raw material. This closed-loop tracking was a proof-of-concept for verifying recycled content. The monetisation aspect involves token rewards (digital badges) to incentivise recycling efforts. The takeaways here extend to carbon management: reciChain shows that *material circularity data* can be reliably captured with blockchain, which in turn provides the data needed to calculate carbon savings (recycled plastic generally has a lower-carbon footprint than newly produced plastic). It also highlights collaboration across a consortium – multiple companies joined the pilot network. A key lesson is that implementing such a system requires aligning stakeholders on data formats and value propositions (why each participant benefits from sharing data). BASF's role as a coordinator was important, suggesting that a strong leading entity can drive adoption of blockchain in a supply chain context. From a carbon perspective, reciChain is a reminder that not all carbon reduction strategies are direct emissions cuts; enabling better recycling cuts carbon indirectly, and blockchain/digital twin tech can quantify and prove those benefits. One challenge observed is scaling beyond a pilot region – the system worked in British Columbia and is expanding to other regions, which will test interoperability and governance on a larger scale.

- **Energy and Climate Blockchain Initiatives:** On a broader, perhaps national level, there are initiatives using blockchain for energy and climate monitoring. For instance, the partnership between Energy-Blockchain Labs (China) and IBM created a platform for China's carbon trading market. While not a digital twin of a single asset, it functions as a *digital twin of the national carbon market*, reflecting companies' emissions and credit trading in real-time on a ledger. The success reported was that it allowed high-emission organisations to monitor their footprints and trade credits with low emitters transparently. Another example is **Power Ledger** in Australia (and other countries) – they implemented blockchain for peer-to-peer energy trading, effectively creating a digital marketplace for solar generation, with some projects also attaching carbon reporting to those transactions. These cases underscore how trusted data exchange can unlock new climate solutions (like peer-to-peer trading or streamlined carbon markets). A takeaway from these is the importance of regulatory support; e.g., the Chinese pilot had backing from a national carbon market context, which likely facilitated acceptance. It also shows that technology is ahead of policy in some areas – these case studies often start as pilots waiting for regulatory frameworks to catch up (for example, recognising blockchain-based credits officially).

In sum, the real-world implementations demonstrate that blockchain and digital twin integration is not just theoretical. Companies are achieving tangible improvements: more accurate emission tracking (Kongsberg, Hyundai), streamlined compliance (Kongsberg, Hyundai), incentive alignment (BASF), and entirely new business models for sustainability (energy trading, carbon markets). The key takeaways include the critical role of cross-sector collaboration (technology providers, industry players, possibly government), the need for performance and scalability in the chosen blockchain platform, and the necessity of clear value propositions for all participants. These case studies also reveal challenges: integration with legacy systems (e.g., linking a factory's digital twin to blockchain required some IT development), ensuring data quality at the source (blockchain secures data *after* input, so IoT sensor calibration and twin accuracy are still vital), and managing the change (training personnel, adjusting processes to rely on smart contracts or blockchain records). Nevertheless, the success stories indicate that these hurdles can be overcome and that the convergence of these technologies can lead to more effective carbon management practices.

## 7.5 Challenges and Future Prospects

While the integration of blockchain and digital twins for carbon tracking is promising, it faces several challenges that must be addressed as well as opportunities for future advancement:

### *7.5.1 Scalability and Performance*

One of the foremost technical challenges is scalability of blockchain networks. Digital twin systems can generate massive amounts of data (e.g., IoT sensors streaming readings every few seconds). Recording every data point on a blockchain is neither practical nor necessary; typically, only aggregated or critical data would be stored. Even so, as the number of transactions grows, a blockchain can become a bottleneck if it cannot handle high throughput.

Public blockchains such as Ethereum historically suffered from limited transactions per second and high fees, though newer solutions (e.g., proof-of-stake networks, Layer 2 scaling) have improved this. In enterprise or consortium blockchains, performance can be tuned, but there is still a limit. As one study on blockchain in carbon trading markets noted, *"a primary challenge is scalability; as transaction volumes increase, the blockchain network can become slower and less efficient,"* which is especially problematic for large-scale carbon markets or IoT-heavy environments (Zhang et al., 2024). If a national digital twin for emissions involved thousands of devices sending updates, the blockchain must scale accordingly. This is driving interest in more efficient consensus algorithms (moving away from energy-intensive proof-of-work to scalable proof-of-stake or BFT algorithms), as well as off-chain solutions (storing bulky sensor data off-chain, and only posting hashes on-chain).

**Prospects:** We expect continued improvement in blockchain scalability through technological innovations such as sharding (parallelising the processing of transactions) and interoperability between blockchains (allowing load distribution). Additionally, not every use case requires global decentralisation – many carbon tracking systems will be on permissioned blockchains (private networks) which can be optimised for higher performance. There is active research into lightweight DLT specifically for IoT, which could directly feed digital twins.

### *7.5.2 Energy Consumption of Blockchain*

There is an irony in using blockchain for climate purposes given the notoriety of early blockchains (like Bitcoin) for high energy usage. A blockchain's carbon footprint depends on its consensus mechanism. Bitcoin's proof-of-work, for instance, consumes more electricity than some countries, undermining climate goals. However, most climate-focused applications use more efficient methods. Ethereum's recent shift to proof-of-stake cut its energy use by >99%. Many enterprise blockchains use consensus algorithms that are not energy intensive. Nonetheless, it remains important to ensure the chosen blockchain platform aligns with the sustainability mission.

**Prospect:** We'll likely see *green blockchain* initiatives (some already underway) – either by using renewable energy to power nodes or developing carbon-neutral consensus. In the future, blockchain networks might routinely publish their own carbon footprint, allowing users to factor that into design. The use of **permissioned networks** for digital twin applications means the network can run on just a few nodes operated by stakeholders, avoiding huge energy waste. The challenge is more applicable to public networks if used for broader transparency.

### *7.5.3 Data Privacy and Ownership*

With increased data sharing on blockchains, there are concerns around sensitive information. Carbon data might be less personal than other data, but for companies, emissions data can be commercially sensitive until they choose to disclose it. Digital twins might also contain proprietary operational insights. Putting all data on a transparent ledger could expose trade secrets or raise privacy issues (in cases where data can be linked to individuals, e.g., digital twins of personal energy usage or transport habits). Regulations like GDPR in Europe impose strict rules about personal data sharing, even if for environmental purposes. A challenge is designing systems where granular data can be used internally by the twin, but only summarised or anonymised data goes on-chain.

**Prospects:** Techniques like *zero-knowledge proofs* are emerging which could allow verification of claims (like "Company A met its carbon target this month") without revealing all underlying data on a public blockchain. Self-sovereign identity (SSI) solutions might give devices or companies control over what data they share and sign off on a blockchain, aligning with privacy by design (St.Clair, 2021). In the future, we might see standardised data governance frameworks for national digital twin networks, where blockchain is used in a privacy-preserving way (for example, only hashes of data or proofs are stored, not raw data).

### 7.5.4 Integration and Interoperability

Many organisations have legacy systems for data collection and reporting. Integrating a blockchain and digital twin solution requires bridging these systems. Interoperability is needed on two levels: between the digital twin software and the blockchain platform, and between different blockchains or data standards if multiple platforms are used. Lack of adopted data standards can hinder the exchange required for digital twins (McKendrick, 2024). For instance, if each company uses a different format for reporting carbon data, simply putting them on a blockchain doesn't solve the reconciliation problem. Also, what if different consortia choose different blockchain solutions? A supplier might face one client on Hyperledger Fabric and another on Ethereum, needing to report to both.

**Prospect:** There is likely to be consolidation or bridges that allow data to flow between platforms. Initiatives like the Digital Twin Consortium and standards bodies (ISO, W3C) are beginning to address interoperability of digital twin data formats. On the blockchain side, projects like Polkadot, Cosmos aim to connect blockchains. Governments could also step in to endorse common standards for national reporting (e.g., a standard token format for a carbon credit, or a standard schema for an emissions report on chain). Overcoming interoperability challenges will make the ecosystem more plug-and-play, so that companies can adopt these technologies without massive custom integration effort.

### 7.5.5 Regulatory and Compliance Concerns

The regulatory environment for both digital twin data sharing and blockchain is evolving. On the blockchain front, issues include the legal recognition of smart contract transactions and tokens, and compliance with financial regulations if tokens are traded (some carbon tokens might be considered securities or commodities, which brings oversight). For digital twins and data, there may be regulations on data residency (keeping data within certain jurisdictions) which conflict with a borderless blockchain network. Moreover, carbon accounting and credits are subject to standards and regulations – blockchain-based methods need to be accepted by regulators and standard setters. A current challenge is that policy often lags technology. Many environmental regulations do not yet contemplate blockchain. However, this is changing: governments are starting to support blockchain for climate action (the EU, for example, is promoting blockchain solutions for carbon tracking (European Commission, 2024).

**Prospects:** We expect clearer guidelines and possibly government-backed blockchain platforms for environmental data. For example, a national government might establish a permissioned blockchain network for companies to report emissions, providing a compliant environment (with rules for verification, privacy, etc.). Another aspect is legal smart contracts – ensuring that what the code does (say, deduct a carbon credit from an account) is legally enforceable or recognised. Efforts in the EU via the proposed regulation of crypto-assets (MiCA) and others might provide a framework. Compliance concerns also include auditability – ironically one of blockchain's strengths. Future audits of emissions could increasingly involve checking blockchain records, so auditors will need tools and skills for that. The compliance frameworks will need to validate methodologies: e.g., if a company claims carbon neutrality via blockchain-tracked offsets, regulators will need to ensure no double-counting between this system and traditional registries. As these issues resolve, confidence in blockchain-enabled carbon accounting will grow.

### 7.5.6 Adoption and Cultural Challenges

Introducing blockchain and digital twins requires change management. Stakeholders might be unfamiliar or sceptical of these technologies. Some may see blockchain as too complex or associate it only with cryptocurrencies. Educating and convincing decision-makers of the ROI is a challenge. There is also the issue of talent – implementing these systems needs people with interdisciplinary knowledge (IoT, simulation, blockchain development, data science). Many organisations face a skills gap in these cutting-edge areas.

**Prospect:** As more success stories emerge (as in Section 7.4), adoption is likely to accelerate, and the technologies will become more user-friendly. We may see *abstracted* solutions where users don't even know they are using blockchain – they just see secure data sharing in their software. Government policies and industry pressure (for transparency) may actually mandate certain approaches, effectively pushing adoption. For instance, if a government digital reporting platform is blockchain-based, companies will adopt it not necessarily because they chose to, but because it's the official system. Culturally, the trustless nature of blockchain sometimes meets organisational cultures built on control – it will take a shift to embrace decentralised responsibility. Over time, as younger, tech-savvy professionals enter sustainability and operations roles, comfort with such technologies will increase. The concept of *"trusted data"* as an asset will become more widely appreciated, and organisations might even get competitive advantage from being seen as transparent and tech-forward in their carbon reporting.

**Prospects:** The future for Blockchain in digital twin-based carbon management are expected to yield a very positive change. We foresee advancements such as integration of IoT, AI, digital twins, and blockchain into unified platforms for *autonomous sustainability management* – imagine AI optimising a digital twin for lower emissions and automatically recording improvements to blockchain and trading saved carbon credits. Another trend could be national digital twin infrastructures that embed blockchain at their core, making secure data exchange a default. As climate commitments tighten, there will be a greater demand for verifiable, real-time tracking – something this integration is uniquely suited to deliver. In the long run, one could envision global carbon ledgers where every major emission source, offset, and mitigation action is transparently accounted for (perhaps under UN oversight), essentially a digital twin of the planet's carbon budget underpinned by blockchain. While ambitious, many pieces of that vision are falling into place. Overcoming current limitations through technical innovation, supportive regulation, and industry collaboration will be key to realising the full potential of blockchain-enhanced digital twins in securing a sustainable future.

## 7.6 Blockchain in National Digital Twin Initiatives

Government and industry bodies worldwide are increasingly developing national or regional strategies for digital twins, particularly for infrastructure and urban systems. As these national digital twin (NDT) initiatives progress, blockchain technology is emerging as a vital enabler for *trusted data exchange and collaboration* across sectors. This section provides an overview of how policies, frameworks, and implementations in the UK, EU, USA, and

China are incorporating blockchain or at least addressing the trust and interoperability challenges that blockchain can solve.

### 7.6.1 United Kingdom (UK)

The UK has been at the forefront with its National Digital Twin Programme (NDTP), launched in 2018 in response to a recognition that better data sharing could unlock huge economic and societal benefits in infrastructure (IET and Atkins, 2019). The **Centre for Digital Built Britain (CDBB)** led this effort, introducing the "Gemini Principles" as foundational guidelines for the NDT ecosystem. These principles include values like *Public Good*, *Value Creation*, and importantly *Trust* – underpinned by security, openness, and quality of data. The vision is an "ecosystem of connected digital twins" across energy, water, transportation, and other infrastructure, all able to interoperate via securely shared data. In such a framework, blockchain is highly relevant.

While early documents of the NDTP focus on principles and an Information Management Framework, technology choices are now emerging. According to one commentary, *Britain launched the NDTP using blockchain*, with CDBB running the program to enable government and others to test decisions in virtual space (Saxena, 2024). This suggests that blockchain was considered from the start as a tool to achieve the trust and security aims of the program. Even if not explicitly mandated in policy, the UK's approach implicitly calls for technologies that ensure data is shared in a trustworthy way across organisational boundaries – a role tailor-made for distributed ledger solutions. Indeed, in the UK's pilot projects (for example, a demonstrator digital twin of critical infrastructure involving the Connected Places Catapult and national infrastructure providers), secure data sharing is paramount (LEAP:IN, 2022). These pilots could leverage blockchain or similar to log data exchanges and enforce permissions.

The UK government's emphasis on *standards* and *federation* for the NDT means that they seek a decentralised way to connect data silos. Blockchain offers a federated data-sharing mechanism where each entity retains control but agrees on a single source of truth for shared data. Moving forward, the UK is likely to integrate blockchain in specific use cases, such as construction supply chains or energy grids, as part of its NDT. The Gemini Principle of *Trust* ("must be trustworthy") can be operationalised by blockchain – ensuring security and transparency of twin data. **Policy status:** With CDBB's five-year mission complete as of 2022 and moving into a business-as-usual phase, one can expect upcoming UK digital twin standards to reference distributed ledgers. Already, the UK has a strong blockchain community and government support for DLT in other areas (e.g., financial services); aligning these with digital twins for climate data and infrastructure management is a natural progression.

### 7.6.2 European Union (EU)

The EU is approaching the digital twin concept at a continental scale with initiatives like the European Green Deal's data space and the Destination Earth (DestinE) project – an ambitious plan to develop a "digital twin of the Earth" for climate and environmental simulations (Garske et al., 2024). The EU's focus is on leveraging big data and AI through digital twins to support policy decisions for sustainability.

A prerequisite recognised by EU strategists is a comprehensive, high-quality data ecosystem, which requires a legal and technical framework for data sharing across member states. To this end, the EU has been actively championing *data governance* regulations (i.e., the Data Governance Act, and Data Spaces) and also blockchain and Web3 strategies. The European Commission explicitly acknowledges blockchain as a tool to help fight climate change and improve emissions transparency (European Commission, 2024). In a policy statement, the Commission notes that *"blockchain is a powerful tool that can significantly improve the transparency, accountability and traceability of greenhouse gas emissions"*, enabling accurate and readily available carbon data. The EU envisions using **smart contracts** to better calculate and track carbon footprints across value chains. This dovetails with the concept of digital twins that represent those value chains or systems. For example, a digital twin of a factory in one country and a supply network in another could feed into a European blockchain-based carbon accounting system, ensuring that if a product is exported as "low-carbon", every step was verified.

The EU's promotion of blockchain for climate is seen in funding programs and innovation calls. They are supporting startups and pilot projects that connect companies, governments, and citizens via blockchain for emissions tracking. The EU's emphasis on *decentralisation* in data (to avoid dominance by big tech and to include all stakeholders) aligns with blockchain's architecture. On the NDT front, various EU member states have their own initiatives (e.g., France and Germany working on industry digital twins, Estonia with government services digital twins, etc.), and the EU often tries to harmonise these. **Policies and frameworks:** The EU's forthcoming regulations on digital product passports (for circular economy) and supply chain due diligence may require interoperable data and audit trails – blockchain integrated with digital twin data could meet these needs. The European Blockchain Services Infrastructure (EBSI) is another piece: EBSI is an EU-backed blockchain network intended for cross-border public services (like verifying certificates, identities, etc.). Though not specific to digital twins, a mature EBSI could be used to certify data coming from NDTs (for example, a country's building emissions data could be timestamped on EBSI to prove authenticity when reporting to EU targets). Additionally, Europe's work on **International Data**

**Spaces** – federated, secure data exchange frameworks – is complementary to blockchain. Those data spaces might incorporate DLT for logging data access and ensuring compliance with usage policies.

In summary, the EU is building an ecosystem where digital twins for sustainability (like DestinE for climate or industrial twins for the Green Deal goals) will operate within an architecture that values open data sharing, trust, and privacy. Blockchain is regarded as an enabling technology to achieve these aims, and EU policy is progressively encouraging its use in climate and data collaboration contexts. We can expect future EU directives to explicitly mention blockchain in contexts like carbon border adjustments, where verifying the carbon content of imports may involve digital twin models of production and blockchain-verified records.

### *7.6.3 United States (USA)*

The US does not yet have a single unified national digital twin program akin to the UK's, but various federal agencies and industry consortia are advancing the concept. Agencies such as the Department of Energy (DOE), Department of Defense, and NIST are interested in digital twins for infrastructure resilience, manufacturing, and smart cities (McKendrick, 2024). A recent National Academies report urged federal coordination on digital twins, noting the lack of standards and the importance of cross-sector data sharing. While blockchain is not explicitly front and centre in federal digital twin discussions, it naturally arises when considering data security and interoperability. For instance, NIST has explored trusted IoT networks and has recognised blockchain as a means to ensure data integrity and device identity.

In smart manufacturing, NIST and other bodies have looked at how to securely share sensor data and events – one prototype approach is using blockchain to log critical events in supply chains or production (for quality or compliance), which parallels carbon tracking. The Digital Twin Consortium (DTC), a global industry-driven body headquartered in the US, includes major tech companies and has working groups on standards. The DTC has identified trust and security as key issues; members have experimented with blockchain, especially for things like provenance of data feeding a twin or licensing of models. One example: a concept of *SSI for devices and twins* has been discussed, where each device/twin has a decentralised identity (possibly blockchain-based) to authenticate itself in a network. This is particularly relevant in the US context of critical infrastructure (power grid, etc.), where a zero-trust architecture is desired – every device must prove its identity and integrity. Blockchain can facilitate this by issuing digital credentials anchored on-chain. On the climate side, the US rejoined the Paris Agreement, and the Biden Administration has focused on climate-tech and infrastructure. Projects like Green Proofs for Bitcoin (to certify renewable mining) or Recycling Traceability pilots have US involvement, indicating an openness to blockchain in environmental monitoring. Some city and state-level initiatives use both digital twins and blockchain: for example, states exploring smart grids with blockchain energy trading, or port authorities using digital twins of logistics and blockchain for tracking shipments (which can include fuel usage/emissions data). Policy environment: The US regulatory approach to blockchain is still chiefly about cryptocurrency regulation, but there's growing acknowledgement of non-financial uses.

The DOE has funded research into blockchain for grid management (to handle distributed energy data), implicitly tying into digital representations of the grid. The challenge in the US is often the siloed nature of agencies – one positive move is creating common data standards (like the recent federal data strategy). If the US were to embark on a NDT for, say, the energy system, incorporating blockchain would help bridge data across private utilities, regional grid operators, and federal oversight. We can foresee pilot programs, perhaps under the DOE or EPA, using blockchain for carbon reporting. For instance, a voluntary program where companies report emissions through a blockchain-based platform could complement EPA's inventory.

In absence of top-down mandate, much US progress comes from industry consortia (like DTC) and private sector innovation. Companies like IBM, Microsoft, and Amazon (all active in both digital twins and blockchain) are providing platforms that government or others can adopt. In sum, while the US doesn't have a formal "blockchain in NDT" policy, the ingredients – advanced R&D, strong industrial base, and emerging government interest in trustworthy AI/IoT – create a context where blockchain will likely be adopted within specific national twin projects (e.g., a virtual twin of the electric grid under DOE's grid modernisation could use a blockchain for device certification and carbon attribute tracking of energy). Future US national strategies on climate-tech might explicitly call out blockchain as part of ensuring data verifiability in climate monitoring systems.

### 7.6.4 *China*

China has invested heavily in both digital twin and blockchain technologies as part of its smart infrastructure and digital economy strategies. The concept of a "digital twin city" took hold in China around 2018, when projects like the Xiong'an New Area (a massive smart city development) were designed with digital twinning in mind (ASEAN, 2024). Chinese cities such as Shanghai, Beijing, and others have been developing urban digital twins to simulate traffic, environment, and city services . Concurrently, China has made blockchain a national priority: in 2019, President Xi Jinping famously declared blockchain an important breakthrough, leading to government support for platforms like the Blockchain-based Service Network (BSN). The BSN is an attempt to provide a nationwide (and even global) infrastructure for blockchain applications at low cost. It's telling that the BSN has been positioned to support many use cases, including smart city and public services. According to one source, *China's BSN verifies identities of its 1.4 billion citizens using blockchain* (Saxena, 2024), illustrating the scale at which China deploys DLT.

In the context of NDTs, having a robust blockchain infrastructure could allow cities and provinces to share data in a trusted manner. For example, a digital twin of transportation in Guangzhou and one in Shenzhen could interoperate via a blockchain-based data exchange if they need to coordinate (say for logistics or disaster management). Policy and implementations: China's government has issued guidelines for new smart cities that include data-sharing frameworks, often with mention of "distributed" and "secure" ledgers. Additionally, China launched a national carbon trading market in 2021 (initially covering power plants), and technological solutions are being used to manage it. The IBM/Energy-Blockchain Labs platform for carbon trading is one concrete implementation, albeit a pilot, that showed how companies can use blockchain to meet carbon market compliance (UNFCCC, 2021b). It aligns well with China's need to monitor and enforce carbon targets across thousands of enterprises. We can anticipate that as this national carbon market expands (it's expected to include more industries), China will incorporate blockchain to ensure data integrity and reduce fraud, given the large-scale and potential for misreporting. Another area is supply chain decarbonisation: major Chinese companies (like Alibaba, Tencent) have been exploring blockchain for supply chain traceability. Combined with digital twin models of product life cycles, this can help meet both China's domestic carbon goals and international requirements (e.g., carbon footprint tracking for exports).

A unique aspect in China is that the government often directly steers tech implementation. If blockchain is endorsed in a particular use case (like tracking green power certificates or renewable energy in the grid), it can become standard. For instance, State Grid Corporation – which is developing an "Energy Internet" – has used IoT and could use blockchain for transactive energy; coupling that with a grid digital twin would allow simulating and settling energy trades with carbon attributes in real time. Frameworks: China's Ministry of Industry and IT (MIIT) has discussed integration of new technologies (blockchain, AI, IoT) in manufacturing – a vision of intelligent manufacturing that surely includes digital twin factories with blockchain for data integrity and supply chain integration. City-level digital twin guidelines (like those by CAICT, a Chinese ICT think-tank) have cited examples where blockchain was used in pilot projects (e.g., Kunming's traffic data twin using blockchain to share data between departments).

Overall, China's approach is pragmatic: use blockchain where it improves trust and efficiency, under government oversight. They may develop proprietary or consortium blockchains (aligned with BSN) for government use. The interplay of blockchain with NDTs in China is likely to grow, particularly with China's dual goals of digital leadership and carbon neutrality by 2060. We might see, in a few years, a city like Shanghai operating a comprehensive urban digital twin with an integrated blockchain-based data governance system, where everything from emissions sensors to energy transactions are recorded immutably, providing the government real time trusted analytics to hit local carbon targets.

In all these regions – UK, EU, USA, China – a common realisation is that achieving the promise of national-scale digital twins (for smarter infrastructure and better environmental outcomes) requires *trustworthy data sharing*. Blockchain is increasingly viewed as a means to that end. The extent of adoption varies: the UK and EU are methodically building frameworks that will likely incorporate blockchain principles; the US relies more on industry-driven adoption, which is happening in pockets; China is top-down implementing both twins and blockchains in parallel. International collaboration could also play a role. For example, if multiple countries want to link their digital twins for climate (perhaps under the Paris Agreement's transparency framework), they might use a blockchain-based system to exchange emissions data with high assurance.

To sum up, national digital twin initiatives are poised to benefit from blockchain by providing the infrastructure of trust. Whether it's ensuring that a city's digital sensor data is credible, that companies share data in a competitive yet collaborative environment, or that citizens can trust digital representations of public systems, blockchain offers solutions that conventional databases cannot easily match in multi-stakeholder scenarios. As policies evolve, we expect to see more explicit integration of blockchain in national digital twin roadmaps, making distributed ledgers a backbone for the next generation of digital infrastructure and carbon accountability at scale.

## 7.7 Conclusion

Blockchain technology and digital twins together present a powerful synergy for advancing carbon target management. Digital twins create detailed, dynamic simulations of physical systems – from individual assets to entire cities – providing a platform to monitor and optimise energy use and emissions. Blockchain complements this by adding a layer of trust, security, and automation to the twin's data and processes. Through this integration, organisations can achieve **real-time, verifiable carbon tracking**: emissions data can be collected by digital twins and immutably logged on blockchain ledgers, ensuring accuracy and preventing tampering. Smart contracts can automate carbon accounting, trigger interventions, or facilitate carbon credit transactions, greatly increasing efficiency.

Case studies from industry leaders (Kongsberg's ESG twin, Hyundai's supply chain emissions system, BASF's reciChain, among others) demonstrate that these concepts are not just theoretical – they lead to practical outcomes like improved transparency, reduced reporting costs, and new incentives for emissions reduction. On a larger scale, national and regional initiatives in the UK, EU, USA, and China underscore the strategic importance of trusted data sharing in realising digital twin benefits for society. They indicate that **policy frameworks are aligning** towards more open, but secure, data ecosystems, where blockchain is often the linchpin for trust.

Our findings highlight several key implications for the future of carbon target management. First, as climate commitments intensify, stakeholders will demand greater **credibility in emissions data** – a role that blockchain-backed digital twins are uniquely suited to fill by providing auditability and accountability. This could accelerate adoption of these technologies across industries that are under pressure to decarbonise. Second, we anticipate the emergence of **standardised platforms** (potentially supported by governments or international bodies) for carbon data exchange, possibly built on distributed ledger foundations; participating in such platforms may become a requirement for companies to demonstrate compliance or to access carbon markets.

Third, the integration of advanced analytics and AI with blockchain and digital twins can enable *predictive* and *prescriptive* carbon management – for example, AI might identify the optimal way to reduce emissions in a process, the twin simulates it, and the blockchain smart contract automatically implements incentives or penalties, creating a feedback loop that drives continuous improvement towards targets. Lastly, collaboration is a recurring theme: achieving net-zero and other targets is a collective endeavour, and the technologies discussed help build networks of trust between entities (whether companies, government agencies, or even consumers). We foresee cross-sector collaborations forming "climate data alliances," leveraging shared digital twin models of supply chains or cities with blockchain ensuring fair and reliable participation.

In conclusion, integrating blockchain with digital twin technology represents a **convergence of the digital and the sustainable**. It transforms carbon management from a retrospective, often opaque accounting exercise into a proactive, transparent, and trustworthy system embedded in operations. This structured, data-driven approach can bolster confidence in reported progress, both internally for decision-makers and externally for regulators and the public. As early challenges around scalability, standards, and adoption are addressed, we expect this approach to become mainstream. The ultimate vision is one of *ubiquitous visibility*: a world where every unit of carbon emitted or saved can be tracked in near real-time through its digital twin and certified on a blockchain, thereby enabling more effective governance of our collective carbon footprint. Achieving such a vision will significantly enhance our ability to meet global carbon targets and mitigate climate change. The journey to that future is underway, and it will require ongoing innovation, supportive policy, and broad collaboration – but the groundwork laid by current research and pilot projects gives cause for optimism that these integrated technologies will play a central role in a more sustainable, data-transparent economy.

Chapter 8

# The Future of Blockchain in Construction

## 8.1 Innovations and Emerging Trends

Blockchain is increasingly recognised as a foundational component of construction's digital transformation, offering decentralised and secure systems capable of addressing long-standing industry inefficiencies such as payment delays, fragmented supply chains, and limited stakeholder trust. At the forefront of innovation are synergies between blockchain, BIM, digital twins, AI, with DAOs beginning to influence governance structures and decision-making processes within the built environment.

Smart contracts, underpinned by blockchain, enable the automation of payment processes, contractual obligations, and workflow execution – significantly reducing administrative overhead and disputes. The integration of digital twins ensures real-time, data-rich representations of physical assets, allowing for performance-based contracts and predictive maintenance regimes.

DAOs are also gaining traction in construction, particularly in governance applications such as collaborative budget allocation, project oversight, and consensus-based decision-making. More advanced iterations – AI-enhanced DAOs – are emerging, employing machine learning to evaluate proposals, optimise resource allocation, and, in some experimental settings, cast autonomous votes on behalf of token holders. These developments signal a shift from passive data registries to intelligent, responsive systems capable of self-regulation and optimisation.

Collectively, these technological convergences signal the emergence of adaptive, resilient, and decentralised construction ecosystems. Yet they also introduce novel design, ethical, and implementation questions that must be critically addressed. This section highlights the accelerating trajectory of blockchain-enabled innovation while framing the necessary groundwork for ensuring that these technologies are meaningfully integrated into the industry's evolving digital infrastructure.

DOI: 10.1201/9781003518303-9

## 8.2 Preparing for Future Challenges

While the innovations in Section 8.1 demonstrate blockchain's potential, the road to widespread adoption in construction is not without obstacles. Industry professionals must prepare for a range of future challenges associated with implementing blockchain solutions. These challenges span regulatory and legal uncertainties, the need for common data standards, and the establishment of governance and risk management practices tailored to distributed ledgers. This section addresses how the construction industry can proactively tackle these issues. In particular, we discuss (1) compliance automation and evolving regulatory frameworks, (2) data standards and interoperability for blockchain adoption, and (3) governance models and risk management strategies. By anticipating these challenges, stakeholders can develop informed strategies to ensure that blockchain implementations are successful, scalable, and aligned with industry norms and laws.

### *8.2.1 Compliance Automation and Regulatory Frameworks*

The introduction of blockchain and smart contracts into construction raises important questions about legal enforceability, regulatory compliance, and industry standards. Compliance automation refers to using technology (in this case blockchain) to automatically monitor and enforce compliance with laws, regulations, and contract terms. For example, a smart contract might be programmed to only execute a payment if a contractor's work is certified to meet building code requirements, effectively embedding regulatory compliance checks into the workflow. This could greatly streamline processes like building inspections, permit issuance, and safety compliance, which today involve significant paperwork and manual oversight.

However, achieving true compliance automation via blockchain requires alignment with regulatory frameworks. Construction is a highly regulated sector – from zoning laws and building codes to labour regulations and environmental standards – and these regulations vary by jurisdiction. One challenge is that current laws did not anticipate blockchain-based processes. For instance, many jurisdictions require wet signatures or notarisation for certain documents, using a smart contract or digital signature on a blockchain may or may not fulfil those requirements legally. There are positive developments: some countries and states have begun recognising blockchain records and smart contracts in legislation (e.g., recognising that a smart contract can constitute a binding agreement), which paves the way for broader use in industries like construction.

Automating compliance on blockchain can also improve accuracy and auditability. Instead of relying on intermittent audits, a blockchain system could provide continuous compliance verification. Zhuk (2025) discusses how regulatory challenges like data privacy, security, and legal recognition are key issues to address for successful blockchain deployment (Zhuk, 2025). In construction, sensitive data (building plans, proprietary BIM data, or personal data of workers) might end up on a blockchain, so compliance with data protection laws (such as GDPR in Europe) is critical. Solutions like permissioned blockchains – where only authorised parties can access certain data – and techniques for off-chain storage of sensitive information with on-chain hashes for verification are being considered to meet privacy regulations.

Moreover, industry regulators themselves may leverage blockchain for compliance monitoring. A building inspector or a funding authority could be a node on a project's blockchain network, enabling real-time oversight. For example, if environmental regulations require documentation of waste disposal from a construction site, a blockchain system could automatically log waste transport receipts and make them visible to regulators, reducing the need for site visits or manual reports. This concept aligns with ideas of "regtech" (regulatory technology) in construction, where compliance checks are integrated into digital platforms. Perera et al. (2020) suggest that future building code compliance might be partly automated: a design in BIM could be run through rule-checking software, and the results (pass/fail for each code clause) could be sealed on a blockchain, providing a verifiable record that the design complies with regulations. A smart contract might even halt progress to the next project phase if compliance verification is missing, thus enforcing a "compliance gate" automatically.

The regulatory frameworks governing blockchain in construction are still evolving. Organisations such as the Construction Blockchain Consortium (CBC) and standards bodies are engaging with policymakers to establish clear guidelines. Key issues under discussion include: How to legally recognise records stored on a blockchain (for contracts, liens, etc.)? How to tax and account for cryptocurrency transactions in projects? How to ensure smart contracts adhere to contract law and do not accidentally violate regulations? There is a consensus that multi-stakeholder collaboration is needed – involving technologists, lawyers, engineers, and regulators – to update contract templates, procurement rules, and standards (e.g., BIM standards or project management protocols) to accommodate blockchain.

One promising development is the concept of "code is law" being tempered with "law is code" – essentially, translating legal and regulatory requirements into machine-readable rules that smart contracts can follow. For instance, a future standard might provide a template smart contract clause for retaining a certain percentage of payment until final inspection is approved, as required by law. By using standardised smart contract libraries that encapsulate regulatory logic, compliance can be automated confidently. However, until such standards mature, early adopters must work closely with legal counsel.

In practice, to prepare for these challenges, construction firms piloting blockchain can start with **sandbox projects** in collaboration with local authorities. This means running a project's transactions on blockchain in parallel with traditional methods, allowing regulators to observe and validate the process. The feedback from such pilots can inform regulators about necessary adjustments to laws or approval processes. It also helps companies develop internal protocols – for example, deciding which compliance documents to put on-chain versus off-chain, and how to handle exceptions when automated rules flag an issue.

In conclusion, while blockchain offers the tools for **compliance automation** (via transparent, immutable records and programmable rules), realising its full benefits will require updates to regulatory frameworks and industry standards. The construction industry should engage proactively with regulators to ensure that smart contracts and blockchain records gain legal acceptance. With clear guidance and perhaps new legislation that accommodates digital ledgers, blockchain can significantly reduce the burden of compliance, making it a continuous, integrated part of the project execution rather than a separate, manual effort. The future likely holds a hybrid approach: human judgement will still govern complex compliance decisions, but blockchain will handle the routine verification and documentation, providing a robust safety net against oversight lapses.

### *8.2.2 Data Standards and Interoperability in Blockchain Adoption*

For blockchain technology to be broadly adopted in construction, it must seamlessly **interoperate** with the industry's existing digital ecosystem. This includes interoperability between different blockchain systems and, crucially, between blockchain platforms and other construction software (such as BIM tools, project management systems, accounting software, etc.). The establishment of **data standards** is therefore a critical challenge. Without common standards, we risk creating isolated "data islands" on blockchains that cannot effectively exchange information with other systems or across different projects and organisations.

In the context of construction, consider the variety of data formats and schemas: BIM models might be in IFC or Revit format, project schedules in Primavera or MS Project, contracts in PDF or proprietary contract management tools, and so on. If a blockchain is to record and reference some of this data, there needs to be a standardised way to represent it on the ledger. For example, if recording a BIM model version on blockchain, using a standard identifier for the model (e.g., an IFC GUID or a checksum of an IFC file) would allow any authorised system to cross-reference that model.

Industry-wide data standards like buildingSMART's IFC for BIM, ISO 19650 for information management, or ISO/TS 23386 for product data, provide a starting point. The challenge is extending these to blockchain or mapping blockchain data structures to them. Singh et al. (2023) highlight that interoperability and standardisation are needed to address technical complexity in blockchain implementation (Singh et al., 2023). They point out that without standards, each new blockchain pilot might develop its own bespoke approach, leading to fragmentation and hampering scalability.

Several initiatives are underway to tackle this. The Construction Blockchain Consortium and other research groups are working on interoperability frameworks that describe how to integrate blockchain with BIM standards and project management standards. One concept is using ontologies – formal representations of knowledge – to bridge between construction domain data and blockchain data. For instance, an ontology can define that a "payment transaction" on a blockchain corresponds to a "payment certificate" in contract administration terms, linking data fields between the systems.

Interoperability also involves the ability of different blockchain networks to communicate. In the future, we might see multiple blockchains in use: one for materials supply chains, another for contract management, another for equipment tracking, etc. If these remain siloed, some benefits of a unified view are lost. Techniques like blockchain oracles (which feed external data into a blockchain) and cross-chain bridges (which allow value or data transfer between blockchains) will be important. For example, a smart contract on one chain could query a delivery status from another chain via an oracle to decide if a penalty clause should execute. Achieving this securely requires standards for cross-chain communication and perhaps the adoption of interoperable blockchain platforms.

The development of open standards for blockchain in construction will likely mirror what happened with BIM: committees of experts from industry and academia will collaborate to define how to structure and exchange data. Already, generic standards are emerging. The World Wide Web Consortium (W3C) has standards for decentralised identifiers (DIDs) and verifiable credentials that could be used in construction to standardise how identities (of companies, or even of products) are represented on a blockchain. For instance, a supplier could have a DID that signs off on the authenticity of a product's data on the blockchain, and any platform understanding W3C's standard would recognise that signature.

Another aspect is ensuring legacy system compatibility. Many construction firms have invested in enterprise resource planning (ERP), procurement, and project management systems. Adoption will be smoother if blockchain solutions offer APIs or plugins that allow these legacy systems to write to and read from the blockchain without massive changes. For example, a procurement system could automatically push purchase order events to the blockchain as transactions, if there is an API mapping the purchase order data to the blockchain's format. This again circles back to standards: a common

API standard for "construction blockchain transactions" would let software vendors build compatible interfaces confidently.

Data standards also influence how governance models work (to be discussed in 8.2.3) because if data is structured and understood uniformly, it's easier to enforce rules and manage permissions. Turk and Klinc (2017) argued that one of the barriers to construction IT adoption is the lack of standardised data flows, a problem which blockchain could exacerbate if not handled carefully. The authors suggested that the goal of interoperability is not total integration but specialisation with communication – meaning each system (or blockchain) does its specialised task but communicates essential info to others reliably (Turk, 2020). In practice, that could mean a supply chain blockchain providing a summarised feed of material status to a project management blockchain in a standardised format.

To prepare for this challenge, industry stakeholders can participate in or follow standardisation efforts such as: ISO committees on blockchain (e.g., ISO/TC 307), CEN/CENELEC initiatives on construction digitisation, and buildingSMART working groups on blockchain and distributed ledgers. Companies can also contribute by documenting their pilot project data structures and sharing what worked, which over time coalesces into best practices.

In conclusion, interoperability and data standards are the glue that will link blockchain into the broader digital tapestry of Construction 4.0. Without them, blockchain applications might remain bespoke solutions with limited impact. With robust standards, however, data can flow across systems – for example, a smart contract could automatically populate a digital twin with performance data, or a BIM model could fetch the latest supply chain status from a blockchain – all in a way that any software vendor or project partner can work with. Achieving this vision requires collective action now to define those standards and ensure that as blockchain adoption grows, it does so on a foundation of interoperability rather than in isolated silos.

## 8.3 The Role of Blockchain in Future Construction Ecosystems

Looking beyond individual use-cases and pilot projects, blockchain is poised to become an integral part of future construction ecosystems. In this final section, we explore a forward-looking perspective on how the industry might evolve with blockchain deeply embedded. We examine three interrelated domains: (1) decentralised project management and collaboration networks, (2) tokenisation and decentralised finance for construction projects, and (3) sustainability and green construction initiatives enabled by blockchain. These topics illustrate the broader impact of blockchain when combined with other technologies and innovative business models. In essence, blockchain can be a catalyst for new ways of organising projects, financing development, and achieving sustainability targets – all key aspects of the construction industry's future.

### *8.3.1 Decentralised Project Management and Collaboration Networks*

Traditional project management in construction relies on centralised coordination by a lead entity (often the general contractor or construction manager) and is documented through siloed systems. In a future blockchain-enabled construction ecosystem, we envision decentralised project management where coordination and information flow occur through a shared network accessible to all project stakeholders. This can be thought of as a project-specific blockchain network that serves as a collaboration backbone – sometimes referred to as a Decentralised Autonomous Project Organisation (DAPO) (Lustenberger et al., 2023). In a DAPO, certain management functions (tracking progress, issuing approvals, updating records) are handled via distributed ledger and smart contracts, rather than a single authority.

What might this look like on a practical level? Imagine each major stakeholder – the client, designers, main contractor, key subcontractors – runs a node in a blockchain network for the project. The project schedule, RFI (Request for Information) logs, change orders, and meeting minutes are all recorded as transactions on this network. Instead of sending separate updates, when a milestone is reached or a task is completed, the responsible party logs it on the blockchain, which all parties' systems subscribe to. Thus, everyone's view of project status is synchronised and trustable. Discrepancies are eliminated because there is one ledger of truth. Decisions can also be recorded – for example, if a change order is approved by a client, their digital sign-off on the blockchain is visible to all, and a smart contract might automatically adjust the project's budget ledger and notify relevant parties.

Decentralisation also opens the door to new collaboration models. For instance, in complex projects, consortiums or joint ventures are formed. A blockchain network can act as the digital JV agreement, allocating shares of decisions and profits via tokens or voting rights encoded on the chain. Husieva et al. (2025) suggest that such models increase transparency in joint projects and ensure each party's contributions and entitlements are fairly recorded. In a way, this is project management by consensus: certain actions require consensus on the blockchain (multiple signatures) which enforces teamwork and reduces unilateral moves that could lead to conflict.

Decentralised project management further implies that knowledge and communications are not bottlenecked. Subcontractors, for instance, often complain they lack visibility into the overall project plan or into design changes that affect their work. With a collaboration blockchain, they can have appropriate access to relevant data (with fine-grained access control ensuring confidentiality where needed). Ye et al. (2024) developed a prototype platform where smart contracts managed construction contract workflows, enabling real-time data exchange and progress tracking among stakeholders (Singh et al., 2023). The result was improved efficiency and transparency in contract management, as each step in the process (submission, review, approval) was visible and automatically logged.

One can also integrate integration with other technologies (IoT) and real-time field data into this network (often termed "Construction 4.0"). For example, imagine drones scanning the site daily and uploading progress data to a blockchain; smart contracts could compare this against the schedule and automatically flag delays or update percent-complete for activities. That information propagates to everyone – the client sees a delay warning, the schedule auto-updates, subcontractors tied to that chain of work get notified of likely start date shifts. In this scenario, blockchain isn't doing project management alone; it's the connective tissue linking IoT, AI analytics, and human decisions in a trusted way.

A decentralised approach may also foster more inclusive decision-making. Typically, subcontractors or suppliers have little say in project execution decisions. But if the project network is designed to solicit inputs (for instance, a smart contract that allows anyone to propose a schedule optimisation which then gets voted on by key stakeholders), good ideas from any source can be recorded and considered. This could democratise some aspects of project problem-solving. It aligns with the broader movement towards integrated project delivery (IPD) and collaborative contracting – blockchain can be seen as a technical enabler of those philosophies, providing the platform to implement shared risk/reward and collective governance.

Of course, decentralised project management doesn't mean no central leadership or accountability; rather it means the data and process backbone is shared. Project managers in the future might spend less time gathering information and more time interpreting the single stream of trusted data and facilitating the best use of it. They become more of "network orchestrators" than information gatekeepers.

A compelling extension of blockchain use in construction lies in scaling beyond individual projects to portfolio- or programme-level management. For example, asset owners such as public infrastructure agencies could consolidate multiple project-specific ledgers into a unified enterprise blockchain, enabling unprecedented levels of oversight, coordination, and data-driven analytics across their entire portfolio. Looking ahead, one can envision smart city infrastructure governed by a network of interconnected blockchains – where each asset or project reports into a city-wide ledger. Within such a system, activities such as maintenance, upgrades, and stakeholder engagement could be coordinated through semi-autonomous processes distributed across the ledger network, significantly enhancing operational efficiency, transparency, and responsiveness.

The role of blockchain in future project management is to enable a decentralised yet coherent network of collaboration, where trust is established by the system's design rather than by one central party. This promises improved transparency, quicker consensus, and a reduction in the administrative burden of reconciling different sources of truth. It aligns well with and enhances existing trends like lean construction and IPD by providing a real-time, trusted information flow. While cultural and contractual shifts are needed to fully realise decentralised collaboration, the technology is progressively demonstrating it can handle the complexity of construction projects at scale. The likely outcome is that tomorrow's construction teams will manage projects not through spreadsheets and isolated systems, but through interconnected digital ledgers that keep everyone literally on the same page (or block).

### *8.3.2 Tokenisation and Decentralised Finance in Construction Projects*

Beyond improving processes, blockchain technology also enables novel financial models for construction through tokenisation and decentralised finance (DeFi). Tokenisation refers to the representation of real assets or financial rights as digital tokens on a blockchain. In construction, this can mean creating digital tokens that represent ownership stakes in a project, claims on future revenue (such as rent or tolls from an infrastructure asset), or even specific assets like pieces of equipment or sections of a building. These tokens can be traded or used as collateral, potentially unlocking new liquidity and democratising investment in the built environment. Decentralised finance extends this concept by enabling financial transactions (like lending, borrowing, insurance, and payment) to occur directly between participants via smart contracts, without traditional intermediaries like banks.

One concrete application is raising capital for projects through tokenisation. Traditionally, large projects are financed via bank loans, bonds, or equity from a few investors. Imagine instead a property developer tokenises a new development project: they issue digital tokens, each representing a small fraction of ownership or entitlement to project cash flows. Investors globally could purchase these tokens, effectively crowdfunding the project but in a regulated, tradable form akin to a security. This is already happening in the real estate sector – for example, in 2019 a $30 million luxury condo in Manhattan was tokenised so that investors could buy shares on a blockchain. For construction, this could mean easier access to financing for developers and an opportunity for smaller investors to participate in infrastructure investments that were previously out of reach. Tian et al. (2022) studied one of the first security token offerings for an energy infrastructure project, finding that tokenisation improved asset liquidity, transaction efficiency, and transparency across intermediaries (Tian et al., 2020). In their case, the token represented an ownership in an energy asset and could be bought or sold on a blockchain marketplace, providing liquidity much sooner than waiting for the project to be completed and sold in entirety.

Tokenisation also allows fractional ownership of construction equipment or facilities. Consider expensive machinery like cranes or tunnelling machines – companies could issue tokens that represent a share of ownership of a machine, enabling multiple stakeholders to invest in it and share its rental income. This could lower the barrier for contractors to access cutting-edge equipment (through fractional leasing models) and for investors to gain returns from construction assets without running a construction firm.

On the DeFi side, decentralised lending platforms could emerge for construction. A contractor with a strong track record recorded on blockchain (e.g., a history of successful project completions and timely payments recorded in smart contracts) could use that as a sort of credit history to secure loans from a decentralised pool of investors. These loans can be governed by smart contracts that automatically repay lenders from project cash flows or trigger penalties on default. For instance, a smart contract might hold a performance bond in escrow and automatically release funds to the client if the contractor fails to meet milestones, providing security to the client while assuring the lender of the project's seriousness. Already, there are experiments in crypto crowdfunding for real estate, where returns are distributed via cryptocurrency dividends to token holders.

Another DeFi concept is stablecoin payments in projects. Instead of transacting in local currency through banks, project stakeholders might agree to use a stablecoin (a cryptocurrency pegged to a fiat currency) on a blockchain. This can speed up transactions (no waiting for bank clears) and reduce forex issues in international projects. Kamel's framework (2023) actually used cryptocurrency (Ether) for executing contract payments (Kamel et al., 2023), showing that it's feasible to handle project funds digitally. As regulatory comfort grows, we might see construction contracts denominated partly in digital currency for efficiency. Some projects in regions with volatile currencies are already eyeing stablecoins as a way to lock in value and pay suppliers overseas with minimal friction.

Decentralised insurance or surety could also play a role. In DeFi, "smart contract insurance" platforms exist that pool funds to automatically pay out when certain conditions are met (like a flight delay insurance paying out if a flight is late, as verified by an oracle). Similarly, in construction, one could imagine an insurance policy for delays or cost overruns that is tokenised. If sensors and ledgers indicate a delay beyond a threshold, a payout could trigger automatically to the affected party. This depends on reliable data feeds, but as IoT and BIM integration increase, it becomes plausible. The benefit would be faster claims resolution and possibly lower premiums, since administrative costs drop.

Of course, tokenisation and DeFi in construction must navigate securities laws and financial regulations. Tokens that represent ownership or debt are usually considered securities in many jurisdictions, so legal compliance (KYC/AML checks on investors, regulatory filings) is necessary. The industry is moving carefully: early tokenised construction financings have been done with heavy legal oversight to ensure they meet existing frameworks (often under crowdfunding or private placement exemptions). Over time, we may see new regulatory categories for digital asset financing of physical projects.

Another forward-thinking idea is incentivising performance and sustainability via tokens. A project could issue its own internal token that is awarded to contractors or workers for hitting targets (like a safety token for each accident-free month, or a green token for recycling materials). These tokens could be redeemed for rewards or trade among participants. It creates a micro-economy within the project that aligns incentives, all tracked on blockchain. While somewhat experimental, it reflects the creativity that token economies allow.

To illustrate the potential magnitude: if major infrastructure projects tokenised 10% of their equity to the public, it could mobilise vast new sources of capital. Public engagement in infrastructure funding might increase if citizens can directly invest in, say, a toll bridge and then automatically receive a small share of toll revenues via smart contract distributions. This blurs the line between traditional public-private partnerships (PPPs) and crowd financing. Discussions on how tokenisation could reshape PPP structures by providing liquidity and new risk-sharing mechanisms has caution that the full potential is yet to be realised due to technical and regulatory hurdles, as well as limited public sector involvement so far. In conclusion, tokenisation and DeFi could unlock new financial frontiers in construction (Tian et al., 2020, 2022). By making assets more liquid and accessible and by automating financial agreements, they promise a construction industry that is more flexible in funding and perhaps less dependent on traditional banking. Stakeholders should watch these developments closely: today's niche pilot (like a tokenised building or a crypto-funded project) could become tomorrow's common practice. As always, it will be crucial to balance innovation with due diligence – ensuring investor protections, project accountability, and alignment with the real-world value that tokens represent. But if done responsibly, blockchain-enabled finance might help bridge the persistent investment gaps in infrastructure and housing by tapping into a global pool of digital-savvy capital.

### *8.3.3 Sustainability and Green Construction Initiatives via Blockchain*

Sustainability is now a paramount concern in construction, with stakeholders across the value chain striving to reduce carbon footprints, improve resource efficiency, and demonstrate compliance with ESG criteria. Blockchain technology offers powerful tools to support green construction initiatives by improving transparency, enabling precise tracking of sustainability metrics, and even creating new incentive mechanisms for environmentally friendly practices. In this section, we explore how blockchain can advance sustainability in construction, from carbon tracking and material reuse to energy efficiency and community engagement in green projects.

One of blockchain's most immediate contributions is ESG data tracking and reporting. Construction projects generate a vast amount of sustainability-related data: carbon emissions from material production and transport, waste generation and recycling rates, energy consumption on-site, labour welfare statistics, etc. Collating and verifying this data for ESG reporting (to clients, investors, or regulators) is often challenging. A blockchain-based system can collect these data points from various sources and lock them into an immutable record. This creates trust in the reported figures and makes it easier to audit sustainability claims. For example, if a contractor claims to use 30% recycled materials, each batch of recycled material can be recorded on blockchain with certifications from suppliers – stakeholders and auditors can then trace exactly how that figure was obtained. Singh et al. (2023) emphasise that blockchain's transparency and immutability can reduce the risk of greenwashing by enabling greater accountability (Singh et al., 2023).

Carbon footprint tracking is a particularly promising area. There are initiatives to attach carbon emission data to construction materials (often called "carbon passports"). By recording these on blockchain, one can sum up the embodied carbon of a building as the materials arrive and are used. If later the building owners want to offset or trade carbon credits, they have a reliable ledger of emissions to reference. Some startups are using blockchain to tokenise carbon credits themselves (each token representing a certified quantity of $CO_2$ saved or offset) (Burgoyne and Olexiuk, 2025). A construction project could automatically acquire and retire such tokens to achieve carbon neutrality, with smart contracts ensuring the credits correspond to the emissions logged on the project's ledger. This kind of integrated approach – tying carbon accounting into the supply chain blockchain – could make real-time carbon management feasible. Instead of waiting until project end to calculate total emissions, project managers could get running totals and even predictions, allowing course corrections (e.g., switching to a greener material) to meet targets.

Another area is materials circularity and green certifications. Blockchain can maintain a material ledger through a building's life, aiding circular economy initiatives. For instance, details of steel beams used (grade, producer, installation date) are on record; decades later, during renovation or demolition, that information can inform whether the beams can be reused or recycled effectively. Shishehgarkhaneh et al. (2023) noted that such traceability provides vital information for decision-making in decommissioning and material reuse. Thus, a building essentially comes with a blockchain-based "materials inventory" that future owners can exploit to minimise waste. Some have coined the term "building material passport", and blockchain is an ideal medium for it because it stays with the asset, independent of any single stakeholder's database.

In terms of operations, blockchain and IoT can combine for sustainable facility management. A digital twin of a building (as discussed in 8.1.2) could monitor energy usage in real-time. Blockchain can record this energy data along with smart contract logic that perhaps interfaces with decentralised energy grids. For example, if a building's solar panels produce excess energy, a smart contract could automatically sell that surplus to a neighbour building or back to the grid, issuing a token or credit in return. The Energy Web Foundation and others are actively exploring such decentralised energy exchanges using blockchain. In a construction project context, say a site office has solar panels or generators – blockchain could manage fuel consumption data and automatically order refuelling when needed, or trade carbon credits for renewable generation, making the site more autonomous and efficient.

Green financing and community incentives can also be implemented via blockchain. A city might issue a "green bond" to finance a sustainable infrastructure project, where bondholders are paid back through smart contracts that draw from project revenue. The blockchain would transparently show how funds are used according to green criteria. On a smaller scale, local communities could be engaged by tokenising certain benefits. Suppose a developer promises community improvements (like planting trees or improving parks) as part of a project's ESG commitments – they could issue a community token that is awarded as these commitments are fulfilled, which residents can see and perhaps trade or redeem in local businesses. While experimental, it demonstrates how blockchain can link the often siloed world of community engagement to tangible project actions via a token economy.

Blockchain can also help with compliance to green building standards (like LEED, BREEAM). Instead of compiling thick binders of documentation to earn certification, relevant data could be continuously gathered on a blockchain throughout construction. By project end, much of the evidence needed for certification is already verified on the ledger – the review process becomes simpler, maybe even automated for certain credits. Auditors could be granted access to the blockchain to perform their evaluation. This feeds back to the compliance automation idea but specifically for sustainability criteria.

A forward-looking possibility is using smart contracts for sustainability incentives: for example, a contract between an owner and contractor might include a bonus for reducing waste below a threshold. Through IoT sensors measuring waste bins and supply chain records of materials, the blockchain system could determine the waste percentage and automatically release the bonus if the target is met. This immediacy and certainty of reward might motivate contractors to implement more sustainable practices rather than view them as extra work.

In all these scenarios, collaboration is key. Achieving sustainability goals often involves multiple parties (contractors, suppliers, waste handlers, energy providers, community groups). Blockchain provides a shared platform where all can contribute and verify data related to sustainability, fostering a more cooperative approach. It aligns well with concepts of shared value – where improvements in environmental performance benefit all by reducing costs or creating new value (like monetisable carbon credits).

Of course, one must acknowledge the energy consumption of blockchain itself, especially public chains (e.g., proof-of-work systems like Bitcoin have high energy use). For construction applications, the use of permissioned or proof-of-stake blockchains, which are far more energy-efficient, is the norm, thus minimising the tech's footprint. In fact, many enterprise blockchains consume negligible energy compared to typical data centres. As blockchain technology evolves (with more eco-friendly consensus algorithms and perhaps even specialised "green" blockchains), this concern will further diminish.

In conclusion, blockchain's role in sustainable construction is as an enabler and enforcer of transparency and accountability. It does not reduce carbon or waste on its own, but it provides the information architecture to ensure sustainability measures are tracked, verified, and rewarded. By doing so, it can strengthen trust in green claims, unlock financial incentives for sustainability (through carbon markets or green investments), and integrate sustainability considerations into the very fabric of project execution through smart contracts. With pressing climate targets looming, the construction industry stands to benefit immensely from such technologies – making the path to net-zero and circular construction not only feasible but verifiable and economically viable.

Overlap analysis reveals numerous studies integrating blockchain with nascent technologies like IoT, AI, and Digital Twin to counteract their individual limitations. As visualised in Figure 8.1, each overlapping section indicates the number of studies focused on a specific technology combination—for example, 11 studies explore blockchain coupled with Digital Twin. Key challenges for blockchain include talent shortages, data privacy, scalability, energy inefficiency, and limited adaptive decision-making. On the right, the diagram outlines how integrating technologies can mitigate these issues: pairing blockchain with IoT decentralises networks and strengthens data security, while combining blockchain and AI enables smart contracts to protect and automate AI model usage. This convergence significantly enhances data security, accountability, and interoperability within construction digitalisation. The figure underscores how Blockchain's synergy with other Construction 4.0

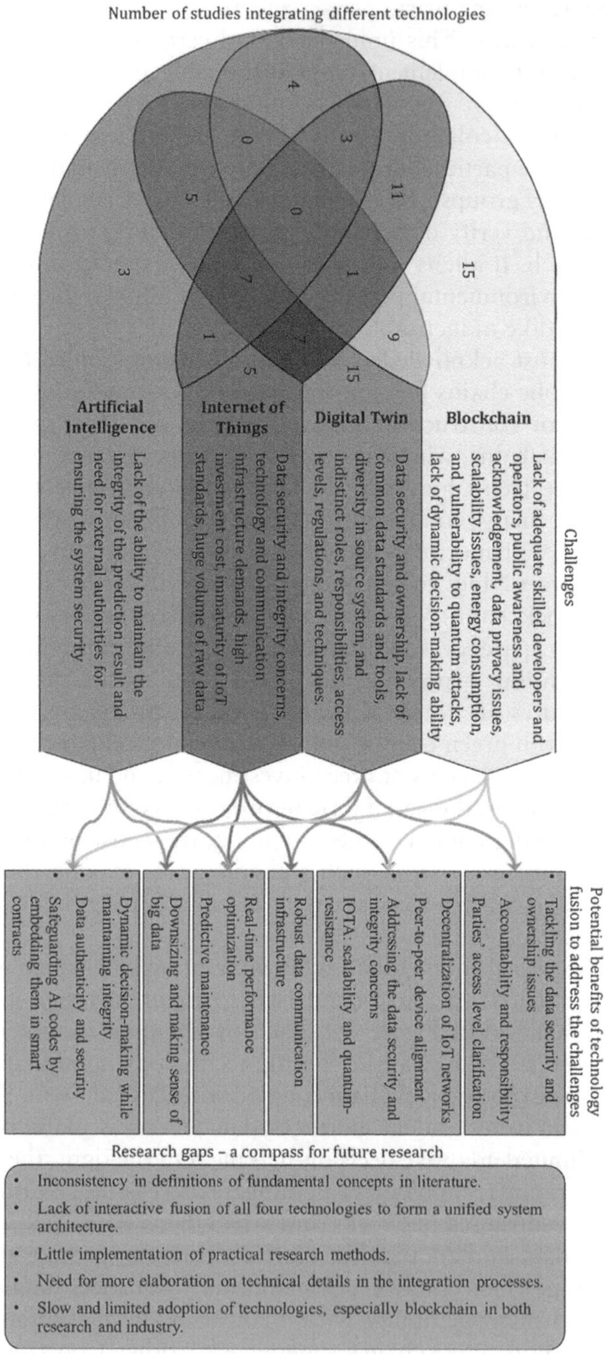


*Figure 8.1* Venn-diagram of blockchain integration challenges.
Source: Sadri et al. (2023).

technologies can build more robust, trustworthy, and sustainable systems for the built environment.

## 8.4 Conclusion

Blockchain technology is charting a transformative course for the construction industry. As detailed in this chapter, its application ranges from improving day-to-day efficiency in contract execution and supply chain management to enabling entirely new models of project organisation and finance. Smart contracts and DAOs promise to automate and secure workflows that were previously prone to delays and disputes, ensuring timely payments and streamlining compliance. The integration of blockchain with BIM and digital twins stands to bolster collaboration and trust in data across a project's lifecycle, addressing long-standing issues of data silos, version control, and cybersecurity in digital engineering. In parallel, supply chain transparency attained through blockchain is not only improving productivity and risk management but also reinforcing ethical and sustainable sourcing by providing an immutable record of materials and products from cradle to grave.

Looking ahead, preparing for wider adoption of blockchain requires tackling several challenges. Regulatory and legal frameworks need to evolve so that blockchain-based processes and records are recognised and supported – the technology must be paired with legal acceptance to reach its full potential. Initiatives around data standards and interoperability are critical to avoid fragmenting the digital landscape; construction stakeholders should coalesce around common standards that allow different systems (and different blockchains) to communicate seamlessly. Effective governance models must be developed to manage the decentralised nature of blockchain networks, ensuring that responsibility and decision-making are clearly defined even when control is distributed. Risk management strategies will need to be updated for the blockchain era, incorporating measures to handle smart contract risks, cybersecurity, and the human factor in technology adoption.

The future construction ecosystem envisioned in this chapter is one where blockchain underpins a more decentralised, democratic, and efficient industry. Project management could become more collaborative and transparent, with blockchain serving as a shared source of truth for all stakeholders. Innovative financing via tokenisation and DeFi may open up construction projects to a broader investor base and provide new ways to manage capital and risk. Sustainability efforts are likely to be strengthened by blockchain through better tracking of emissions, resources, and compliance with green practices, providing credible proof of a project's environmental and social impact. In essence, blockchain can act as a trust infrastructure for Construction 4.0 – enabling trust not just between people, but between digital systems and data streams that drive modern project delivery and operations.

It is important to acknowledge that blockchain is not a panacea or a plug-and-play solution. Its successful implementation in construction depends on thoughtful integration with other technologies like IoT, AI, cloud, BIM, and, crucially, on buy-in from people: industry professionals, firms, regulators, and clients must collaborate to reshape processes and mindsets. The case studies and research discussed show encouraging results, but also highlight that pilot projects often start small – focusing on specific issues like payments or materials tracking – and then scale up. This incremental approach is likely the pragmatic path for the industry: identify high-friction areas where blockchain adds value, demonstrate success, and gradually expand the scope.

In conclusion, the future of blockchain in construction is both promising and purpose driven. The technology aligns with the industry's broader goals of increasing productivity, reducing disputes, enhancing collaboration, and achieving sustainability. As the construction sector continues to digitalise, blockchain provides a foundational layer of trust and automation that can knit together the many facets of digital engineering and project management. For construction professionals, innovators, and digital leaders, embracing blockchain is not about jumping on a hype bandwagon – it is about critically assessing where it can solve real problems and improve outcomes. Those that do so early will help set the standards and best practices for everyone else, positioning themselves at the forefront of an industry revolution. With careful implementation, blockchain can help build not just structures, but stronger partnerships and more resilient processes across the built environment value chain.

Chapter 9

# Construction Enterprises and Projects Adoption Strategies

## 9.1 General Structural Considerations

Embracing blockchain within the built environment requires a holistic view of technical, organisational, and cultural factors. A roadmap must be structured well enough and address how best to scope blockchain solutions, establish effective oversight, create a supportive data ecosystem, secure the right resources, and manage change. Focusing on these key dimensions ensures that enterprises not only integrate distributed ledger technologies smoothly but also seize the tangible benefits of enhanced collaboration, security, and transparency. Whenever possible, drawing from real-world lessons and industry standards helps navigate the technical complexities, legal considerations, and practical challenges. Consider the following:

### 9.1.1 *Approach – Define Scope and Strategic Fit*

Organisations must start by identifying high-impact use cases (e.g. supply chain traceability, smart contracts), aligned with business goals and stakeholder needs. A baseline assessment of current data processes, readiness, and pain points sets the stage. The roadmap should be anchored in regulatory alignment, risk management, and a delivery model that translates vision into measurable outcomes.

### 9.1.2 *Governance – Establish Oversight and Accountability*

Clear governance structures – roles, rules, and responsibilities – are critical for coordinating blockchain activities such as node deployment, consensus mechanisms, and data ownership. Formal processes ensure legal compliance, transparency, and traceable metrics, combining on-chain and off-chain data to demonstrate value and sustain stakeholder trust.

DOI: 10.1201/9781003518303-10

### 9.1.3 *Commons – Create a Shared Operational Foundation*

Success depends on a shared understanding of blockchain terms and their relevance across asset lifecycles. Aligning blockchain standards with existing frameworks (e.g. ISO19650) enables interoperability. Security, data quality, and permissioning must be governed by robust, testable protocols – initially via pilots – to build a repeatable model for scaling.

### 9.1.4 *Enablers – Secure Resources and Build Capability*

Implementation requires financial investment, cybersecurity planning, and international collaboration. Industry leaders must advocate adoption, while skills development ensures operational competence. Lessons from BIM and early pilots can refine approaches and accelerate readiness across the sector.

### 9.1.5 *Change – Manage Culture, Communication, and Adoption*

Effective change management involves clear role definitions, ongoing communication, and engagement with all stakeholders. Explaining blockchain's tangible benefits fosters buy-in. Commercial conditions (e.g. procurement or token models) must be addressed early. Capturing lessons learned supports adaptation and long-term integration into digital practices.

These high-level structural key points present a holistic framework for adopting blockchain in the built environment – bridging technical execution with organisational readiness. By systematically addressing scope, governance, shared infrastructure, enablers, and change, the framework supports a resilient and scalable transition towards more transparent, efficient, and collaborative construction ecosystems.

## 9.2 Roadmap for Blockchain Adoption in Construction

### 9.2.1 *Identify High-Value Business Use Cases*

Blockchain initiatives must begin with specific business pain points – not the technology. There should be a focus on areas where traditional systems struggle due to lack of trust, poor transparency, or the need for multi-party consensus. Relevant examples include payment delays, supply chain opacity, regulatory compliance tracking, and ESG traceability.

Use structured frameworks – such as the World Economic Forum's Blockchain Beyond the Hype decision tree – to assess whether blockchain's core features (immutability, shared write access, auditability) offer a real advantage over conventional databases. Prioritise use cases where blockchain can eliminate frictions like manual verification, fragmented record-keeping, or reconciliation bottlenecks. Define measurable KPIs (e.g. reduction in payment cycle time, decreased administrative overhead, improved material provenance).

### 9.2.2 *Select the Right Blockchain Architecture*

The choice of blockchain model – public, private, or consortium – has major implications for performance, governance, and legal exposure:

- Public blockchains (e.g. Ethereum) maximise transparency and decentralisation, suitable for tokenisation, digital identity, and open innovation but often face regulatory uncertainty and performance constraints.
- Private (permissioned) blockchains offer controlled access and predictable latency, better suited for internal process automation but expose participants to central operator risks.
- Consortium blockchains distribute governance among sector peers, balancing trust and efficiency. These are currently dominant in supply chain, finance, and trade logistics.

Use industry guidance (e.g. Leow, 2024) to match your data-sharing intensity, competition landscape, and stakeholder trust dynamics to the appropriate model. Legal and compliance teams should be engaged early to assess jurisdictional exposure and liability implications of each option.

### 9.2.3 *Pilot Strategically, with Layer 2 and Interoperability in Mind*

Start with limited-scope pilots or proofs-of-concept that test blockchain's technical feasibility and economic benefits in a controlled setting. Select cross-functional teams – IT architects, project managers, legal counsel, finance specialists – and relevant external stakeholders (e.g. contractors, suppliers, platform vendors).

Design these pilots with interoperability and scalability from the outset:

- Use Layer 2 solutions (e.g. roll-ups, sidechains) to address performance and cost issues without compromising on Layer 1 security.
- Plan for cross-chain compatibility using standardised messaging protocols and APIs to avoid vendor lock-in and future-proof the solution architecture.

Track early KPIs rigorously. Use go/no-go gates to halt, refine, or scale up based on real data. Recognise that many blockchain supply chain initiatives remain in pilot stage – use this phase to build internal competence and stakeholder confidence.

### 9.2.4 *Integrate with Enterprise Systems and Data Pipelines*

Integration should be planned from day one to ensure seamless operation across existing platforms such as ERP, procurement, BIM, or asset management systems. Use middleware or APIs to ensure blockchain functions are embedded into user workflows without adding friction.

Ensure that blockchain becomes a source of truth without duplicating data or creating shadow systems. Apply strong cybersecurity and identity management protocols – e.g. integrating the blockchain platform into the organisation's IAM framework to define who can read, write, and validate transactions.

### 9.2.5 *Establish Governance and Compliance Structures*

Blockchain governance must define:

- Roles and responsibilities (e.g. node operators, validators)
- Data ownership, access rights, and dispute resolution mechanisms
- Smart contract approval workflows and update policies

Embed compliance-by-design from the outset. Implement on-chain controls such as KYC filters, permissioning layers, and immutable audit trails. Follow international guidance such as the WEF's Pathways to the Regulation of Crypto-Assets and OECD recommendations on responsible blockchain innovation.

Assign legal and cybersecurity professionals to development teams early to avoid costly retrofits. Maintain a compliance heat map across jurisdictions and update it as regulatory clarity evolves.

### 9.2.6 *Build Organisational Buy-In*

Internal adoption is just as critical as technical success. Share pilot outcomes clearly with all stakeholders – executives, finance teams, operations leads, and legal counsel. Provide training for relevant personnel on how to interact with the blockchain layer (e.g. digital wallets, on-chain record verification). Establish executive sponsorship, steering committees, and cross-functional working groups to manage change.

Address cultural and procedural resistance explicitly. Demonstrate how blockchain initiatives align with broader digital transformation goals, ESG objectives, and competitive differentiation.

### 9.2.7 *Collaborate Through Ecosystem Partnerships*

Blockchain networks thrive on collaboration, not isolation. Explore:

- Industry consortia to share infrastructure costs (e.g. for material passports or trade finance platforms)
- Technology partnerships with Layer 2 providers, oracle services, and custodians to reduce time-to-market
- Open-source engagement to influence protocol direction and attract talent

World Economic Forum guidance emphasises the importance of multi-stakeholder alignment for ESG metrics, sustainability goals, and long-term trustworthiness.

### 9.2.8 *Scale and Evolve the Platform*

Once pilots are validated, broaden the scope incrementally – onboarding more functions, projects, or supply chain partners. Monitor impact through updated KPIs (e.g. payment cycle time, RFI turnaround, regulatory reporting accuracy). Feed insights from blockchain analytics back into strategic planning.

Be ready to evolve the platform. Update governance, protocols, or partners as standards mature and new technologies (e.g. AI-integrated DAOs) become viable. Engage with standard-setting bodies to help shape sector-specific norms – particularly relevant in the built environment, where interoperability and compliance are paramount.

This roadmap synthesises strategic frameworks and operational steps for blockchain adoption in construction and infrastructure sectors. By aligning business goals, architecture decisions, compliance structures, and collaborative ecosystems, organisations can avoid the pitfalls of early blockchain hype and realise tangible, scalable benefits across their portfolios.

## 9.3 Best Practices and Potential Pitfalls

### 9.3.1 *Best Practices*

Successful blockchain adoption in the built environment requires a structured approach that balances technical feasibility, stakeholder collaboration, and regulatory alignment. The following best practices provide a clear roadmap for maximising blockchain's benefits while mitigating common challenges:

1. Start with Simple, High-Impact Applications: Prioritise use cases that offer immediate value and require minimal disruption, such as supply chain tracking, smart contracts for payments, or asset verification. Starting small builds confidence and allows for iterative improvements.
2. Develop a Reliable Data Ontology: Ensure that all stakeholders agree on a standardised data structure for blockchain integration. A shared

ontology enhances interoperability, prevents data fragmentation, and ensures meaningful adoption across disciplines.

3. Incorporate Integration Checkpoints: Establish validation gates within the project lifecycle to assess blockchain feasibility early. These checkpoints help identify technical or regulatory misalignments before large-scale deployment.
4. Prioritise Interoperability Across Systems: Design blockchain solutions that integrate with existing IT infrastructures, legacy databases, and other blockchain networks. Interoperability ensures seamless data exchange and avoids silos.
5. Focus on Incremental Adoption and Scalability: Start with pilot projects and gradually expand. Scalability planning ensures the blockchain solution can accommodate growing data volumes, user participation, and network activity without performance issues.
6. Define Governance and Compliance Strategies Early: Establish clear governance rules, data ownership policies, and compliance frameworks to prevent operational bottlenecks. Ensure blockchain solutions align with industry regulations and standards such as ISO19650 for BIM.
7. Implement Robust Smart Contract Testing: Smart contracts must undergo rigorous testing to prevent security vulnerabilities and unintended outcomes. Use simulations, formal verification, and real-world testing before deployment.
8. Prioritise Privacy and Security: Blockchain solutions must protect sensitive construction data. Implement permissioned blockchains, encryption, zero-knowledge proofs, and multi-signature access controls to safeguard information.
9. Invest in Training and Skills Development: Provide education for developers, project managers, and on-site teams to ensure blockchain solutions are correctly implemented and maintained. Knowledge gaps can hinder adoption and efficiency.
10. Leverage Tokenisation for Asset Management: Tokenising construction assets, from equipment to project milestones, enables better tracking, fractional ownership, and automated auditing. This can improve accountability and resource utilisation.
11. Integrate Off-Chain Data Thoughtfully: Not all construction data needs to be stored on-chain. Use off-chain oracles and hybrid models to balance blockchain's immutability with external data requirements, ensuring efficiency and cost-effectiveness.
12. Engage Industry Leaders and Standards Bodies: Collaborate with blockchain consortia, industry working groups, and regulatory bodies to align solutions with best practices, foster interoperability, and ensure widespread adoption.

13. Monitor, Maintain, and Iterate: Blockchain is not a "set-and-forget" technology. Continuous monitoring, performance audits, and periodic updates are necessary to address emerging risks, improve efficiency, and adapt to evolving industry needs.

### 9.3.2 *Pitfalls on Blockchain Technology Adoption*

While blockchain offers transformative potential for the built environment, its integration is fraught with pitfalls, many of which stem from technological bottlenecks. Recognising these challenges early can help organisations navigate them effectively and avoid project derailment. Below are key pitfalls to consider:

1. **Over-engineering Early Solutions**: A common mistake is designing overly complex blockchain systems during initial implementation. Advanced features, such as multi-chain interoperability or AI-enabled smart contracts, can overwhelm teams and delay timelines. Without starting with simpler, proven applications (e.g., supply chain tracking or smart contracts for payments), organisations risk stagnation and wasted resources.
   *Example Pitfall*: Implementing a full-scale blockchain solution for asset lifecycle management without validating smaller applications may result in poorly integrated systems and unforeseen incompatibilities.
2. **Poorly Defined Ontology and Data Standards**: Without a well-defined and shared ontology, blockchain systems risk becoming fragmented and unscalable. Inconsistent data structures, lack of uniform semantics, and incompatible metadata can hinder interoperability and undermine blockchain's ability to serve as a single source of truth.
   *Example Pitfall*: In construction projects, BIM data that is not mapped to a consistent blockchain-friendly schema may result in incomplete or inaccurate records, disrupting workflows reliant on data consistency.
3. **Lack of Scalability Testing**: Failure to assess blockchain's scalability under project demands can lead to performance bottlenecks. Public blockchains may struggle with high transaction volumes, while private ones might suffer from limited node capacity or inefficient consensus mechanisms.
   *Example Pitfall*: A materials tracking system integrated into a blockchain may perform well during pilot projects but fail to handle real-time data from multiple suppliers in a full-scale project.
4. **Ignoring Integration Challenges in Legacy Systems**: Blockchain must coexist with existing IT and operational systems. Ignoring integration challenges – such as incompatible APIs, outdated infrastructure, or

on-premises data silos – can lead to fragmented operations and increased technical debt.
*Example Pitfall*: A project management blockchain solution that fails to sync with existing ERP systems might create duplicate data entry requirements, reducing efficiency rather than improving it.

5. **Misaligned Expectations for Real-Time Processing**: Blockchains, especially public ones, can suffer from latency issues due to consensus mechanisms and transaction finality. Stakeholders expecting real-time processing may become frustrated when systems cannot deliver immediate results.
*Example Pitfall*: Delays in processing payments via smart contracts due to network congestion can strain supplier relationships and project timelines.
6. **Underestimating Security Vulnerabilities**: Blockchain's security depends on its implementation. Weak encryption practices, poorly written smart contracts, and insecure key management can expose systems to attacks or fraud. Additionally, relying on off-chain components may introduce vulnerabilities.
*Example Pitfall*: Exploitable smart contract bugs can lead to financial losses, as seen in high-profile cases like decentralised finance (DeFi) hacks, highlighting the need for rigorous testing.
7. **Over-reliance on Blockchain for Inappropriate Use Cases**: Blockchain is not a universal solution. Applying it to processes where simpler, more cost-effective technologies suffice can waste resources and diminish its perceived value.
*Example Pitfall*: Using blockchain for routine document storage instead of focusing on its strengths, such as immutable audit trails or automated contract execution, can lead to unnecessary complexity.
8. **Neglecting Interoperability Across Multiple Chains**: In multi-stakeholder environments, blockchain platforms often differ in terms of protocols, data structures, and governance. Neglecting interoperability can isolate project ecosystems, limiting collaboration and creating inefficiencies.
*Example Pitfall*: A contractor using a private blockchain may find it difficult to share data securely with subcontractors operating on public blockchains without intermediary solutions.
9. **Insufficient Planning for Network Governance**: Failure to define governance mechanisms – such as who controls nodes, approves changes, or resolves disputes – can result in operational deadlocks or network inefficiencies.
*Example Pitfall*: Disagreements among stakeholders on governance issues, such as data ownership or validation processes, can lead to mistrust and adoption delays.

10. **Resource and Energy Demands of Blockchain Networks:** Some blockchain systems, particularly those relying on proof-of-work, are resource-intensive and environmentally unsustainable. Failing to account for these energy demands can conflict with sustainability goals in construction.
    *Example Pitfall*: Using energy-intensive blockchains for carbon tracking might undermine green building certifications and client expectations for sustainable solutions.

## 9.4 Training and Skill Development for the Future Workforce

Training is a critical yet often overlooked component of successful blockchain or any technological adoption in the construction sector. To build robust organisational capacity and lower adoption barriers, structured programs must be prioritised – focusing on core concepts, tools, and practices. Hands-on, face-to-face training embedded within real-world project settings is particularly effective, as it promotes experiential learning and fosters peer-to-peer knowledge exchange. Furthermore, integrating training into agile delivery processes enables iterative skill development, rapid feedback, and practical alignment with evolving project demands. This approach ensures that blockchain competencies are not only developed but continuously refined in pace with implementation.

## 9.5 Conclusion

Blockchain adoption within the construction sector is not a question of if, but how effectively and sustainably it can be implemented. This chapter has laid out a comprehensive strategy for integrating distributed ledger technologies into enterprise operations and project workflows – starting with a structured roadmap, reinforced by best practices, and supported through continuous workforce development.

Key to success is the deliberate alignment of blockchain capabilities with real-world project challenges – prioritising transparency, trust, and automation where traditional systems fall short. Yet, as clearly illustrated through the identified pitfalls, technological enthusiasm must be tempered with practical caution. Poor integration, governance ambiguity, or over-engineered solutions can derail progress and erode value.

To ensure meaningful outcomes, organisations must invest not only in infrastructure and technical frameworks but also in change management, stakeholder collaboration, and skills development. Agile, well-informed teams – supported by strong governance and grounded in industry standards – will be central to scaling blockchain solutions across the built environment.

Ultimately, the path forward is incremental yet strategic: starting small, building confidence through pilot success, and scaling with purpose. When approached holistically, blockchain offers the construction industry a rare opportunity to reshape its digital foundation – delivering a more secure, efficient, and collaborative future.

Chapter 10

# Legal and Ethical Considerations

## 10.1 Navigating the Legal Landscape

The "Panopticon effect" originates from Jeremy Bentham's 18th-century architectural design of the Panopticon – a circular prison in which inmates could be observed by a central guard without knowing when they were being watched (Bentham, 1795, 1776). The power of this design (see Figure 10.1 and Figure 10.2) lies not in the act of surveillance itself, but in the possibility of being watched, which induces individuals to self-regulate their behaviour. This dynamic – the perception of constant oversight producing voluntary compliance – has found renewed relevance beyond its penal origins.

In economic theory, particularly within mechanism design, the Panopticon effect offers a means to structure multi-party environments where self-interested actors are incentivised to act in socially or organisationally desirable ways. By fostering the belief that deviations from the rules may be detected—even if not continuously monitored – the costs of enforcement can be significantly reduced. This principle underpins many real-world systems, from financial regulation and corporate governance to workplace management and online reputation systems, where strategic uncertainty about oversight drives compliant behaviour and minimises the need for intrusive surveillance.

Blockchain technology elevates this logic to a new plane. Through cryptographic transparency, immutable ledgers, and automated rule enforcement via smart contracts, blockchain enables a form of decentralised observation in which actions are permanently recorded and verifiable by all participants. The result is a technologically mediated Panopticon – dispersed, layered, and often obfuscated – where behaviour is shaped not by visible guards but by the structural inevitability of auditability and traceability. In such environments, trust is engineered through code, and compliance becomes the default, not the exception.

Yet, this technological amplification of the Panopticon effect also raises pressing ethical concerns – especially when entwined with artificial intelligence. As AI systems grow more autonomous and predictive, they blur the

DOI: 10.1201/9781003518303-11

*Figure 10.1* Box 119 – Panopticon Sectional Elevation.

Source: Bentham (1795).

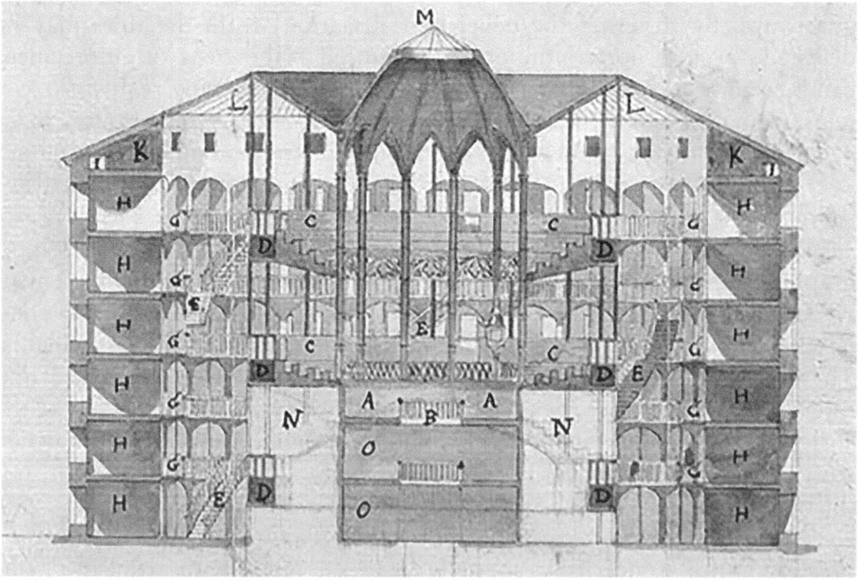


*Figure 10.2* Panopticon Cross-Section.

Source: Reveley (1791).

boundaries between monitoring and control, raising questions about consent, accountability, and the distribution of power (Coeckelbergh, 2020). The challenge is to harness these tools to promote accountability and cooperation without replicating oppressive surveillance logics. In navigating this new legal and ethical terrain, the construction and governance of digital infrastructures must be guided as much by normative reflection as by technical sophistication.

## 10.2 Current Legislation Worldwide

Four major jurisdictions are compared regarding how they classify and regulate cryptoassets. Although each aims to improve consumer protection and prevent financial crimes, their methods diverge:

### *10.2.1 United Kingdom (UK)*

The UK's overarching goal is to integrate cryptoasset regulation into existing financial frameworks, avoiding any imbalance between crypto businesses and traditional institutions (Xiong and Luo, 2024). In 2018, the Cryptoassets Taskforce (CATF) – comprising the Financial Conduct Authority (FCA), Bank of England, and HM Treasury – was formed (Treasury, 2018). CATF advocated decisive measures to mitigate cryptoasset risks and maintain the UK's position as a leading global financial hub (UK GOV, 2017).

In 2019, the FCA announced that cryptocurrencies could be regulated as securities if they fall under the Financial Services and Markets Act (FSMA) 2000 (Regulated Activities) Order 2001 (FCA, 2019) The Financial Services and Markets Act, 2000 Regulated Activities Order, 2001a; 2001b). Subsequently, in 2023, HM Treasury released "Future Financial Services Regulatory Regime for Cryptoassets", outlining plans to gradually integrate cryptoassets into the FSMA framework. This consultation noted that exchange tokens (e.g., Bitcoin, Litecoin) and utility tokens (e.g., Ethereum) would be subject to the Payment Service Regulations 2017 if used for regulated payment services. Shortly thereafter, the FCA mandated all cryptoasset firms, including overseas entities, to register for compliance with the UK's financial promotions regime. This requirement extends the same consumer protection found in traditional financial services.

A landmark ruling, AA v Persons Unknown, confirmed exchange tokens may be regarded as property (Low, 2020; Vos, 2019). Drawing on National Provincial Bank v Ainsworth (Holland and McRae, 1965), the court concluded that assets like Bitcoin can be considered property. While security tokens are treated as securities, and exchange tokens may be classified as property, utility tokens are still unregulated, neither recognized as securities nor property.

Thus, most cryptoassets in the UK are governed as either securities or property, based on their characteristics and uses. Given the minimal discussion in policy and literature, the UK does not treat cryptocurrency as a commodity. However, by embedding cryptoassets into the same framework as fiat and traditional banking, the UK gradually extends comparable consumer protection and financial crime prevention measures to the crypto sphere.

### *10.2.2 United States of America (US)*

Like the UK, the US seeks to integrate cryptoasset regulation into existing frameworks. The Commodity Futures Trading Commission (CFTC) treats Bitcoin as a commodity (Lucking and Aravind, 2020) under the Commodities Exchange Act (U.S. Government, 1936), following an enforcement action that deemed Bitcoin to fit the broad definition of a commodity (CFTC, 2015). The Internal Revenue Service (IRS) regards Bitcoin as property (IRS, 2025),recently reinforced by final regulations requiring custodial brokers to report digital asset sales. Meanwhile, the Securities and Exchange Commission (SEC) considers certain cryptocurrencies securities under the Securities Exchange Act (SEC, 1975), guided by the Howey test (Johnson, 1984). This has enabled high-profile enforcement actions, such as in the Ripple case and the Kim Kardashian endorsement fine (SEC, 2024).

Because the CFTC, SEC, and IRS each take different stances, the US regulatory landscape remains fragmented, lacking a single cohesive framework. This ambiguity hinders consumer protection and financial crime prevention, as the applicable safeguards depend heavily on which classification – commodity, security, or property – is used. While these agencies aim to protect consumers and curb illicit activity, the current patchwork approach can undermine those goals by creating uncertainty about how cryptocurrencies should be regulated.

### *10.2.3 Japan*

Japan integrates crypto regulation into existing legislation, treating cryptocurrencies as securities under the Financial Instruments and Exchange Act (JLT, 2010) or as legal property under the Payment Services Act (2019). Determining whether a token is a security or property depends on its characteristics (Library of Congress, 2022). This approach arises from three legal reforms (Amaya, 2022):

- **First Legal Reform (2016)** – Triggered by the G7 Elmau Summit (EC, 2022), FATF guidance (FATF, 2024a), and the Mt Gox collapse (Rao, 2021). Japan amended the Payment Services Act and Act (PSA, 2019) on Prevention of Transfer of Criminal Proceeds (FATF, 2024a) to strengthen consumer protection, including mandatory registration for crypto exchanges(Omagari and Sako, 2019).

- **Second Legal Reform (2019)** – Addressed speculative trading and new models like ICOs, partly spurred by the Coincheck hack in 2018 (Tsuchiya and Hiramoto, 2021). Japan's two crypto trade groups merged into the self-regulatory Japanese Cryptocurrency Exchange Association, overseen by the Financial Services Association (FSA) (Goto et al., 2019). Further amendments to the Payment Services Act and Financial Instruments and Exchange Act (1948) introduced rules on crypto derivatives, advertising, and solicitation (Amaya, 2022).
- **Third Legal Reform (2022)** – Focused on stablecoins, following international agreements (Financial Stability Board recommendations) and global growth in stablecoin use. Amendments to the Banking Act (1981), Payment Services Act, and Trust Business Act (2004) established frameworks for banks, trust companies, and fund transfer providers to issue stablecoins, and required identity verification for intermediaries.

Overall, Japan's integration of crypto into existing laws emphasizes financial stability and consumer protection – similar to the UK's framework but with stricter measures following incidents like Mt Gox and Coincheck. Stablecoins fall under the Banking Act, Payment Services Act, and Trust Business Act, security tokens under the Financial Instruments and Exchange Act, and other tokens under the Payment Services Act (Amaya, 2022).

### *10.2.4 China*

China stands out with its unique method of cryptoasset regulation. In 2013, the People's Bank of China (PBC) issued the "Notice on Preventing Bitcoin Risk" (PBC, 2021), declaring that Bitcoin lacks the same legal status as the Renminbi (RMB) and is merely a "specific virtual commodity." This notice aimed to safeguard the RMB's legal tender status, reduce money laundering risks, and preserve financial stability. It also required Bitcoin trading platforms to register with telecom regulators under the Telecommunications Regulations of the People's Republic of China (2020) and the Measures for the Administration of Internet Information Services. Like Japan, it mandated that institutions fulfil AML obligations in line with the Anti-Money Laundering Law of the People's Republic of China (PBC, 2016) .

With rising Initial Coin Offerings (ICOs), the PBC released the "Announcement on Preventing Risks of Token Issuance Financing" (PBC, 2017), labelling ICOs as unauthorized fundraising. It banned using and exchanging "virtual currencies" like Bitcoin and Ethereum, criminalizing any virtual asset service providers. Later, the "Notice on Further Preventing and Handling the Risk of Speculation in Virtual Currency Transactions" (PBC, 2021) reconfirmed China's stance, declaring crypto-related business activities – —including mining – illegal.

China's ban aligns with global goals of consumer protection and financial crime prevention but diverges sharply by outlawing all crypto activities. While it treats cryptocurrencies as a virtual commodity, the US partially does so only for Bitcoin via the CFTC, while the IRS and SEC use different classifications. Japan and the UK instead integrate crypto into existing frameworks, though the UK's approach is less clear than Japan's formal distinctions for stablecoins, security tokens, and payment tokens.

## 10.3 Regulating Cryptoassets on An International and Regional Level

When looking at cryptocurrency regulations, it is vital to consider the varying approaches of different countries and how international and regional organizations attempt to harmonize these frameworks. Given crypto's borderless nature, a coordinated approach is key to addressing its unique challenges while maximizing its benefits.

### *10.3.1 Financial Action Task Force (FATF)*

As an intergovernmental body combatting money laundering and terrorist financing, the FATF (Mantalara, 2021) sets international standards for Anti-Money Laundering (AML) and Counter-Terrorist Financing (CTF), urging legal, regulatory, and operational measures against illicit finance. Since 2014, it has focused on virtual assets, issuing specific Recommendations for industry participants and national supervisors.

**Recommendation 15:** Addresses risks from new technologies and virtual assets, requiring countries and financial institutions to assess money laundering and terrorist financing threats (FATF, 2024b). It highlights how anonymity, decentralization, and global reach make cryptocurrencies attractive for criminal activities.

**Recommendation 16:** Governs wire transfers, mandating accurate originator and beneficiary data throughout the payment chain. Institutions must monitor transfers lacking required information, freeze suspicious assets, and enforce United Nations Security Council resolutions. Although non-binding, the FATF can use sanctions to push adoption, making certain measures (e.g., Recommendation 16) effectively mandatory (Alexander, 2001).

The FATF is pivotal in shaping AML/CTF norms worldwide for cryptocurrencies. Through Recommendations 15 and 16, it stresses risk assessment, accurate transaction data, and the prevention of misuse by criminals—bolstering financial security, international cohesion, and consumer protection across global markets.

### 10.3.2 European Union (EU)

The EU, a political and economic union of 27 countries, seeks to harmonize Member States' approaches to crypto regulation by creating Directives and Regulations that integrate cryptoassets into existing domestic frameworks. In 2018, it introduced the Fifth Anti-Money Laundering Directive (EU, 2018), which mandated "Know Your Customer" (KYC) measures for virtual exchanges to address concerns over illicit crypto use. However, because each Member State can interpret directives independently, regulations remained uneven across the EU.

To resolve this, the EU adopted the Markets in Crypto Assets (MiCA) Regulation in May 2023 (CSSF, 2025). MiCA's goal is to boost financial stability and enhance consumer protection, covering asset-reference tokens, E-money tokens, and utility tokens, yet excluding cryptoassets already regulated under other financial legislation (e.g., security tokens via MiFID II (EU, 2014). It also requires cryptoasset issuers and service providers to obtain authorization from national authorities.

Like the UK and Japan, the EU initially integrated crypto regulation into existing frameworks, then recognized the need for harmonization, leading to MiCA's unified structure. This ensures consumer protection, financial crime prevention, and continued innovation. Meanwhile, the FATF—through Recommendations 15 and 16—urges global standards on AML/CTF, risk assessment, and transparency. Although non-binding, these recommendations significantly influence worldwide adoption. With domestic and international regulation frameworks addressed, the next important development is the potential role of Central Bank Digital Currencies (CBDCs) in shaping future cryptocurrency regulation.

### 10.3.3 The Forefront of Regulation

There is no single universal consensus on the leading country for cryptoasset regulation, as several jurisdictions have implemented advanced or comprehensive frameworks tailored to their specific markets. However, Switzerland and Singapore frequently emerge as top contenders in discussions about forward-thinking, robust, and business-friendly regulatory environments for digital assets.

*Switzerland*

- **Crypto Valley in Zug:** Switzerland has fostered a thriving blockchain ecosystem known as "Crypto Valley" in the canton of Zug. Here, blockchain startups benefit from clear and welcoming regulations, low taxes, and well-established legal structures.

- **FINMA's Regulatory Approach:** The Swiss Financial Market Supervisory Authority (FINMA, 2025) provides detailed guidelines for issuing and trading tokens, including ICO/STO (Initial Coin Offering / Security Token Offering) rules. It classifies tokens as payment, utility, or asset (securities) tokens, offering legal clarity for businesses.
- **Distributed Ledger Technology (DLT) Legislation:** In 2021, Switzerland updated its legal framework (often referred to simply as the "DLT Act") to recognize ledger-based securities and ease the tokenization process. This legislation further cements Switzerland's reputation for being pro-innovation while maintaining strong investor protections.

Switzerland's balanced stance—marked by a general openness to innovation coupled with rigorous oversight—has made it an attractive destination for blockchain initiatives, establishing it as a leading European hub for crypto-currency and Decentralised Finance (DeFi).

*Singapore*

**Payment Services Act (PSA):** Enforced by the Monetary Authority of Singapore (MAS), the PSA regulates payment systems and digital payment tokens. Crypto exchanges, custodians, and other service providers must obtain relevant licenses, comply with Anti–Money Laundering (AML) and Counter Terrorism Financing (CTF) rules, and follow consumer protection requirements.

**Regulatory Flexibility:** MAS has proven adaptive, issuing consultation papers and regular updates. It has also encouraged fintech innovation while tightening regulations on areas like retail access to crypto, especially after market volatility and high-profile collapses.

**Strong Fintech Ecosystem:** Singapore's pro-business climate, robust infrastructure, and government grants have made it a hotbed for fintech development, attracting diverse crypto and blockchain startups.

Singapore's clear licensing regime and strong oversight balance innovation with risk mitigation. Its approach signals that crypto-friendly policies do not preclude rigorous consumer protection and anti-illicit finance measures.

Chapter 11

# Conclusion

Blockchain technology has evolved rapidly over the past two decades, maturing from a niche cryptographic innovation underpinning Bitcoin into a foundational digital infrastructure with far-reaching implications. In the context of construction – a sector historically marked by fragmented processes, adversarial contracting practices, and significant productivity gaps – blockchain presents both a compelling opportunity and a formidable challenge. This book has traced the contours of blockchain's relevance to the built environment, examining its theoretical foundations, cross-sectoral applications, and transformative potential when combined with adjacent technologies such as building information modelling (BIM), digital twins, artificial intelligence (AI), and the Internet of Things (IoT).

As the construction industry navigates the complexities of Construction 4.0 and begins to envisage the contours of Construction 5.0, blockchain emerges not merely as a technological innovation but also as a catalyst for rethinking how projects are initiated, financed, delivered, and maintained. However, this promise remains partially realised, and its realisation depends not only on technical feasibility but also on organisational alignment, institutional readiness, and regulatory coherence.

## 11.1 Towards Decentralised, Transparent, and Trustworthy Construction Systems

One of the most consistent themes throughout this book has been blockchain's ability to establish *trust in trustless environments*. The construction sector, which is replete with information asymmetries, misaligned incentives, and contractual disputes, stands to benefit substantially from decentralised record-keeping, immutable data structures, and automated verification mechanisms.

Smart contracts are perhaps the most visible and immediate application of blockchain in construction. By enabling the conditional execution of predefined clauses – such as releasing payments upon milestone verification – they directly address common sources of friction, such as delayed payments,

DOI: 10.1201/9781003518303-12

performance disputes, and opaque subcontracting chains. When linked to BIM models and IoT sensor data, these contracts enable dynamic, verifiable, and real-time workflows, potentially reducing administrative burden and increasing accountability. Chapter 3's case study on Chinese wage disbursement mechanisms illustrates how this principle, when implemented at scale, can also support social inclusion, labour transparency, and project efficiency.

Beyond smart contracts, tokenisation unlocks new frontiers in construction financing. By fractionalising ownership of assets or future revenue streams, blockchain enables more liquid, inclusive, and transparent capital structures. This model supports the emergence of decentralised finance (DeFi) mechanisms and decentralised autonomous organisations (DAOs), which have been discussed in Chapter 5. These structures offer alternative models for project governance, enabling distributed stakeholder engagement in budgetary oversight, procurement decisions, and even design inputs.

Chapter 6 demonstrated how blockchain can act as a federating layer across BIM and CDE ecosystems, preserving data provenance and streamlining audit trails. In a sector where claims and counterclaims often revolve around the question of "who knew what and when," blockchain's immutable ledger can serve as a neutral arbiter. This is particularly relevant in complex, multistakeholder environments, such as PPP infrastructure projects, international construction ventures, or integrated project delivery (IPD) frameworks.

Sustainability is now a nonnegotiable imperative. Chapter 7 explored how blockchain-enabled digital twins can support robust carbon tracking, emission monitoring, and compliance with climate targets, particularly in the context of national net zero ambitions. By embedding trusted carbon data into the construction lifecycle and potentially tokenising verified carbon reductions, blockchain offers a powerful compliance and incentive mechanism for sustainable development. The implications for environmental, social and governance (ESG) reporting and green finance are substantial.

As Chapter 8 makes clear, blockchain is likely to form a foundational layer of future construction ecosystems—alongside AI, digital twins, and advanced robotics. Whether through traceable material passports, automated permits, or resilient infrastructure registries, blockchain can help to future-proof the sector against increasingly volatile economic, environmental, and geopolitical risks in the future.

## 11.2 Between Technological Potential and Institutional Reality

However, enthusiasm for blockchain must be tempered by a realistic appraisal of the persistent—and in some cases, structural—barriers to its widespread adoption. While pilot projects are proliferating and conceptual interest is high, the translation of potential into practice remains sporadic and uneven.

Despite the appeal of blockchain principles, their implementation is far from straightforward. The sector faces a multitude of blockchain platforms (e.g., Ethereum, Hyperledger, Corda), each with varying architectures, scalability profiles, and governance models. Chapter 1 explained that selecting an appropriate consensus mechanism or data model entails significant technical trade-offs in terms of throughput, privacy, cost, and energy usage. The absence of a dominant standard complicates interoperability, particularly across jurisdictional or organisational boundaries.

Moreover, many construction enterprises lack the internal expertise to evaluate these choices, leading to vendor lock-in, failed pilots, or inappropriate use cases. Misunderstanding around blockchain's capabilities—confusing it with simple databases or assuming "magic-bullet" benefits—can also lead to disillusionment or poor resource allocation.

Blockchain adoption is not merely a technical endeavour—it demands organisational change. As Chapter 9 discussed, firms must develop strategic roadmaps, prepare their workforce, and align blockchain applications with core business objectives. However, entrenched habits, hierarchical project structures, and a risk-averse culture often stymie innovation. The construction industry's historically slow rate of digitalisation is a known issue, and blockchain adoption risks repeating the BIM implementation gap—where tools are deployed but not embedded in processes or culture.

Furthermore, the decentralising ethos of blockchain runs counter to the centralised project management paradigms common in large-scale construction. Moving from top-down control to distributed consensus requires a rethinking of governance, liability, and accountability—all of which remain underexplored in mainstream project management literature.

Legal frameworks for blockchain are evolving but unevenly. As Chapter 10 highlights, global jurisdictions differ radically in their treatment of cryptocurrencies, smart contracts, and digital assets. The EU's Markets in Crypto-Assets (MiCA) regulation, Japan's staged reforms, and China's prohibitionist stance represent divergent policy logics – balancing innovation, risk, and control. For multinational construction firms or cross-border projects, this regulatory heterogeneity poses real barriers.

Equally problematic is the lack of clear legal recognition of smart contracts, particularly in common law jurisdictions. Until such contracts are deemed enforceable and the role of oracles (as discussed in Chapter 4) is legally validated, the use of automated decision-making in high-stakes construction projects will remain limited to low-risk or experimental domains.

While blockchain promises transparency, it can also introduce new forms of surveillance and data centralisation under the guise of decentralisation. Chapter 10 discusses the "Panopticon effect", whereby actors are continuously monitored, not by a single authority but by the architecture of the system itself. This may reinforce accountability but also erodes privacy and autonomy – particularly for subcontractors, temporary workers, or marginalised groups.

Furthermore, the convergence of blockchain with AI and the IoT raises complex ethical and algorithmic governance questions: who sets the rules encoded in smart contracts? How is bias identified in training data? What recourse exists when decisions go wrong or when automation fails? These nascent issues must be addressed proactively through interdisciplinary collaboration among technologists, lawyers, ethicists, and construction professionals.

## 11.3 Strategic Imperatives for Construction Stakeholders

Against this backdrop of opportunity and challenge, several imperatives emerge for industry stakeholders wishing to engage strategically with blockchain.

Successful blockchain adoption requires more than technological literacy. It demands institutional frameworks for data governance, liability allocation, and dispute resolution. Construction enterprises should collaborate with regulators, insurers, and legal experts to define new standards for smart contract enforceability, blockchain auditability, and tokenised asset recognition.

Blockchain should not be deployed indiscriminately. Its strengths – immutability, transparency, and decentralisation – make it ideal for certain applications (e.g., supply chain tracking, milestone verification, and IP management) but overkill for others. Firms should adopt an incremental, use-case-driven approach, focusing on demonstrable ROI and learning-by-doing through controlled pilots.

Chapter 9 rightly emphasises the need for upskilling. Blockchain literacy must be embedded across project management and legal, commercial, and IT functions – not confined to a digital innovation team. Cross-training professionals in both construction workflows and blockchain architecture will be crucial. Partnerships with academic institutions, industry bodies, and certification schemes can support this effort.

The value of blockchain increases with the number of interoperable participants. Isolated implementations offer limited returns. Instead, consortium-based models – industry platforms for procurement, permitting, or asset management – should be prioritised. Examples from other sectors (e.g., trade finance or pharmaceutical traceability) offer blueprints for multistakeholder governance.

Finally, blockchain's long-term impact will hinge on our capacity to anticipate its unintended consequences. This includes planning for data permanence (e.g., what happens to outdated or erroneous records? ), avoiding technological monocultures and ensuring that automation does not displace human judgment or exacerbate inequality. Ethical foresight must be embedded into innovation pathways – not bolted on after deployment.

## 11.4 Final Thoughts

Blockchain represents a paradigmatic shift in how information, value, and trust are organised and exchanged. In the construction sector – where disputes are common, transparency is rare, and efficiency gains are long overdue – its relevance is profound. However, like any foundational technology, blockchain adoption is neither inevitable nor uniformly beneficial. The sector must engage with it critically, strategically, and ethically.

This book has offered a conceptual, empirical, and practical grounding in blockchain's application to construction. It is intended not as a definitive statement but as an informed starting point for professionals, researchers, and policymakers seeking to navigate the blockchain frontier. The construction sector is standing on the brink of digital reformation; whether it harnesses blockchain to build more equitable, efficient, and sustainable futures depends on the actions taken today.

# References


A3Logics, 2025. Blockchain for Construction: A Guide to Use Cases, Benefits, Challenges and Solutions. A3Logics. URL www.a3logics.com/blog/blockchain-for-construction/ (accessed 6.1.25).

Alexander, K., 2001. The International Anti-Money-Laundering Regime: The Role of the Financial Action Task Force. *J. Money Laund. Control* 4, 231–248. https://doi.org/10.1108/eb027276

Allaire, J., 2024. Blockchain: In from the Cold and Set to Disrupt the World of Finance [WWW Document]. World Economic Forum. URL www.weforum.org/stories/2024/01/blockchain-change-world-finance-stablecoins-internet/ (accessed 6.1.25).

Almeida, A.D., Vrees, R., 2020. Empowering Participation in the Energy Transition with Blockchain. IBM, 28 July 2020, www.ibm.com/think/insights/empowering-participation-in-the-energy-transition-with-blockchain#:~:text=A%20decentralized%20and%20democratic%20system,essential%20in%20addressing%20these%20issues

AlShamsi, M., Al-Emran, M., Shaalan, K., 2022. A Systematic Review on Blockchain Adoption. *Appl. Sci.* 12, 4245. https://doi.org/10.3390/app12094245

Amaya, T., 2022. A Speech on "Regulating the Crypto Assets Landscape in Japan" at the Crypto Asset Roundtable Hosted by OMFIF [WWW Document]. URL www.fsa.go.jp/en/news/2022/20221207/20221207.html (accessed 7.26.25).

Asean, C.R., 2024. Digital Twins: A New Model for Urban Governance [WWW Document]. URL www.prnewswire.com/apac/news-releases/digital-twins-a-new-model-for-urban-governance-302081193.html (accessed 3.24.25).

Azhar, S., Nadeem, A., Y N Mok, johnny, H Y Leung, B., 2008. Building Information Modeling (BIM): A New Paradigm for Visual Interactive Modeling and Simulation for Construction Projects.

Badhwar, A., Islam, S., Tan, C.S.L., 2023. Exploring the Potential of Blockchain Technology Within the Fashion and Textile Supply Chain with a Focus on Traceability, Transparency, and Product Authenticity: A Systematic Review. *Front. Blockchain* 6. https://doi.org/10.3389/fbloc.2023.1044723

BASF, 2023. Viewpoint on PLASTICS RECYCLING [WWW Document]. URL https://plastics-rubber.basf.com/global/en/plastics-hub/BASF_views_on_plastics/viewpoint-on-plastics-recycling (accessed 7.20.25).

Bender, M.C., 2022. DeFi: An Opportunity or an Existential Threat to Nanks? – Financial Services Blog [WWW Document]. deloitte.com. URL www.deloitte.com/au/en/Industries/financial-services/blogs/defi-opportunity-existential-threat-banks.html (accessed 6.1.25).

Bentham, J., 1795. The Panopticon [WWW Document]. UCL Museums and Collections. URL https://ucl.primo.exlibrisgroup.com/permalink/44UCL_INST/5apqbq/alma9931777345004761 (accessed 3.5.25).

Bentham, J., 1776. *The Correspondence of Jeremy Bentham*. 1, https://uclpress.co.uk/book/the-correspondence-of-jeremy-bentham-volume-1/ (accessed 13.12.24).

Benton, M.C., Radziwill, N.M., Purritano, A.W., Gerhart, C.J., 2018. Blockchain for Supply Chain: Improving Transparency and Efficiency Simultaneously. Software Quality Professional 20, 28.

Burgoyne, M.T., Olexiuk, P., 2025. Tokenized Carbon Credits: How Blockchain is Revolutionizing Carbon Markets. Osler Hoskin Harcourt LLP. URL www.osler.com/en/insights/updates/tokenized-carbon-credits-blockchain-revolutionizing-markets/ (accessed 7.20.25).

Buterin, V., Wood, G., 2013. Ethereum Whitepaper [WWW Document]. URL https://static.peng37.com/ethereum_whitepaper_laptop_3.pdf (accessed 1.21.25).

Cai, L., Sun, Y., Zheng, Z., Xiao, J., Qiu, W., 2021. Blockchain in China – Communications of the ACM. URL https://cacm.acm.org/research/blockchain-in-china/ (accessed 6.1.25).

Casino, F., Dasaklis, T.K., Patsakis, C., 2019. A Systematic Literature Review of Blockchain-based Applications: Current Status, Classification and Open Issues. *Telemat. Inform.* 36, 55–81. https://doi.org/10.1016/j.tele.2018.11.006

CBC, 2021. White Papers [WWW Document]. CBC. URL www.constructionblockchain.org/white-papers (accessed 6.14.21).

Cecere, L., n.d. Tradelens Discontinues Operations. Why You Should Care. [WWW Document]. Forbes. URL www.forbes.com/sites/loracecere/2022/12/05/tradelens-discontinues-operations-why-you-should-care/ (accessed 9.24.24).

CFTC, 2015. Order Instituting Proceedings Pursuant to Sections 6(c) and 6(d) of the Commodity Exchange Act, Making Findings and Imposing Remedial Sanctions.

Chainlink, 2020. Chainlink: The Industry-Standard Oracle Platform [WWW Document]. URL https://chain.link/ (accessed 11.21.25).

Chaum, D., 2024. Home [WWW Document]. chaum.com. URL https://chaum.com/ (accessed 12.13.24).

Chaum, D., 1982. Blind Signatures for Untraceable Payment. *Adv. Cryptol.* 82, 199–203.

Cheng, M., Liu, G., Xu, Y., Chi, M., 2021. When Blockchain Meets the AEC Industry: Present Status, Benefits, Challenges, and Future Research Opportunities. *Buildings* 11, 340. https://doi.org/10.3390/buildings11080340

CM, 2019. Vinci and Skanska blockchain tool 'delivers 25% cost savings.' Construction Management. URL https://constructionmanagement.co.uk/vinci-and-skanska-blockchain-tool-delivers-25-cost/ (accessed 6.1.25).

Coeckelbergh, M., 2020. *AI Ethics*. The MIT Press.

ConExpo, 2025. Unlocking Efficiency and Trust with Smart Contracts in Construction [WWW Document]. Conex.-CONAGG. URL www.conexpoconagg.com/news/unlocking-efficiency-and-trust-with-smart-contract (accessed 7.26.25).

Construction Management, 2019. Vinci and Skanska Blockchain Tool 'Delivers 25% Cost Savings' [WWW Document]. Construction Management. URL https://constructionmanagement.co.uk/vinci-and-skanska-blockchain-tool-delivers-25-cost/ (accessed 5.4.25).

Cozzi, L., Gould, T., 2019. What Would It Take to Limit the Global Temperature Rise to 1.5°C? – Analysis [WWW Document]. IEA. URL www.iea.org/commentaries/what-would-it-take-to-limit-the-global-temperature-rise-to-15c (accessed 3.6.25).

CSSF, 2025. Markets in Crypto-Assets (MiCA/MiCAR) – CSSF [WWW Document]. URL www.cssf.lu/en/markets-in-crypto-assets-mica-micar/ (accessed 7.26.25).

Dai, J., Vasarhelyi, M.A., 2017. Towards Blockchain-Based Accounting and Assurance. *J. Inf. Syst.* 31, 5–21.

Davis, F.D., 1989. Perceived Usefulness, Perceived Ease of Use, and User Acceptance of Information Technology. *MIS Q.* 13, 319–340. https://doi.org/10.2307/249008

Davis, T., Henry, W., Hansen, B., Massey, R., 2024. Tokenization: Realizing the Vision of a Future Financial Ecosystem. April 2024, Zotero, 01/06/2025.

Develle, Y., 2024. European Blockchain Sandbox Announces the Selected Projects for the First Cohort – EBSI SANDBOXCollab – [WWW Document]. URL https://ec.europa.eu/digital-building-blocks/sites/display/EBSISANDCOLLAB/European+Blockchain+Sandbox+announces+the+selected+projects+for+the+first+cohort (accessed 6.1.25).

Develle, Y., Thomas, J., 2024. European Blockchain Sandbox Announces the Selected Projects for the First Cohort – EBSI SANDBOXCollab – [WWW Document]. URL https://ec.europa.eu/digital-building-blocks/sites/display/EBSISANDCOLLAB/European+Blockchain+Sandbox+announces+the+selected+projects+for+the+first+cohort (accessed 6.1.25).

EC, 2022. G7 Summit, Schloss Elmau [WWW Document]. Consilium. URL www.consilium.europa.eu/en/meetings/international-summit/2022/06/26-28/ (accessed 7.26.25).

Ekman, A., 2021. China's Blockchain and Cryptocurrency Ambitions | European Union Institute for Security Studies [WWW Document]. URL www.iss.europa.eu/publications/briefs/chinas-blockchain-and-cryptocurrency-ambitions (accessed 7.26.25).

Esmaeilian, B., Sarkis, J., Lewis, K., Behdad, S., 2020. Blockchain for the Future of Sustainable Supply Chain Management in Industry 4.0. *Resour. Conserv. Recycl.* 163, 105064. https://doi.org/10.1016/j.resconrec.2020.105064

EU, 2018. Directive – 2018/843 – EN – aml directive – EUR-Lex [WWW Document]. URL https://eur-lex.europa.eu/eli/dir/2018/843/oj/eng (accessed 7.26.25).

EU, 2014. Directive – 2014/65 – EN – mifid ii – EUR-Lex [WWW Document]. URL https://eur-lex.europa.eu/eli/dir/2014/65/oj/eng (accessed 7.26.25).

European Commission, 2024. Blockchain for Climate Action | Shaping Europe's Digital Future [WWW Document]. Blockchain Clim. Action Eur. Comm. Aims Use Innov. Blockchain Technol. Help Fight Clim. Change. URL https://digital-strategy.ec.europa.eu/en/policies/blockchain-climate-action (accessed 3.6.25).

Evans, S., Savian, C., Burns, A., Cooper, C., 2019. Digital Twins for the Built Environment: An Iintroduction to the Oopportunities, Bbenefits, Cchallenges and Rrisks. 17 Oct 2019, Zotero, IET.

FATF, 2024a. Japan's Progress in Strengthening Measures to Tackle Money Laundering and Terrorist Financing [WWW Document]. URL www.fatf-gafi.org/content/fatf-gafi/en/publications/Mutualevaluations/japan-fur-2024.html (accessed 7.26.25).

FATF, 2024b. The FATF Recommendations [WWW Document]. URL www.fatf-gafi.org/en/publications/Fatfrecommendations/Fatf-recommendations.html (accessed 7.26.25).

FCA, 2019. Cryptoassets: Our Work [WWW Document]. FCA. URL www.fca.org.uk/firms/cryptoassets (accessed 7.24.25).

FDA, 2024. Center for Devices and Radiological Health, 2024. Center for Devices and Radiological Health: Radio Frequency Identification (RFID). FDA.

FINMA, 2025. Welcome to the Swiss Financial Market Supervisory Authority FINMA | FINMA [WWW Document]. URL www.finma.ch/en/ (accessed 7.26.25).

Fitriawijaya, A., Taysheng, J., 2024. Empowering Digital Twin Through BIM–Blockchain for Carbon Disclosure of Certified Green Buildings. *Comput.-Aided Des. Appl.* 180–202. https://doi.org/10.14733/cadaps.2025.180-202

Garske, B., Holz, W., Ekardt, F., 2024. Digital Twins in Sustainable Transition: Exploring the Role of EU Data Governance. *Front. Res. Metr. Anal.* 9. https://doi.org/10.3389/frma.2024.1303024

George Lawton, 2025. Top 8 Blockchain Platforms to Consider in 2025 | Informa TechTarget [WWW Document]. Search CIO. URL www.techtarget.com/searchcio/feature/Top-9-blockchain-platforms-to-consider (accessed 5.4.25).

Goto, I., Kuribayashi, Y., Saito, T., 2019. Virtual Currencies: Regulators Make Their Move. *Int. Financ. Law Rev.* 2019, 75.

Grilo, A., Jardim-Goncalves, R., 2010. Value proposition on interoperability of BIM and collaborative working environments. Automation in Construction, Building Information Modeling and Collaborative Working Environments 19, 522–530. https://doi.org/10.1016/j.autcon.2009.11.003

Hackitt, J., 2018. *Building a Safer Future: Independent Review of the Building Regulations and Fire Safety*. Dandy Booksellers Limited.

Hansen, M., 2022. ESG Delivery: A Punch of Power with Digital Twins and Blockchain [WWW Document]. Kongsberg. URL www.kongsbergdigital.com/resources/esg-delivery-a-punch-of-power-with-digital-twins-and-blockchain (accessed 3.6.25).

Hassan, S., De Filippi, P., 2021. Decentralized Autonomous Organization. *Internet Policy Rev.* 10 , 1–10.

Hatzivasilis, G., Ioannidis, S., Fysarakis, K., Spanoudakis, G., Papadakis, N., 2021. The Green Blockchains of Circular Economy. *Electronics* 10, 2008. https://doi.org/10.3390/electronics10162008

Hijazi, A.A., Perera, S., Calheiros, R.N., Alashwal, A., 2022. A data model for integrating BIM and blockchain to enable a single source of truth for the construction supply chain data delivery. *Engineering, Construction and Architectural Management* 30, 4645–4664. https://doi.org/10.1108/ECAM-03-2022-0209

Holland, D.C., McRae, D.M., 1965. National Provincial Bank v. Ainsworth [I9651 AC 1175, is re-examined.

Hong, K., 2016. Bitcoin as an Alternative Investment Vehicle. *Inf. Technol. Manag.* 18, 265–275.

Howcroft, E., 2022. NFT Sales Hit $25 Billion in 2021, but Growth Shows Signs of Slowing. Reuters.

Husieva, Y., Chumachenko, I., Skachkov, O., Kadykova, I., Dotsenko, N., 2025. Using Blockchain Technology for Monitoring Projects, in: Lytvynov, O., Pavlikov, V., Krytskyi, D. (Eds.), 2025. Using Blockchain Technology for Monitoring Projects. *Integrated Computer Technologies in Mechanical Engineering* – 2024. Springer Nature Switzerland, Cham, pp. 461–472. https://doi.org/10.1007/978-3-031-94845-9_38

HYUNDAI, 2023. Hyundai Motor and Kia to Introduce Blockchain-based Carbon Emission Monitoring System to Foster a Sustainable Value Chain [WWW Document]. HYUNDAI Mot. URL www.hyundai.com/worldwide/en/newsroom/detail/hyundai-motor-and-kia-to-introduce-blockchain-based-carbon-emission-monitoring-system-to-foster-a-sustainable-value-chain-0000000280 (accessed 3.24.25).

IBM, 2021. What Is a Digital Twin? [WWW Document]. URL www.ibm.com/think/topics/what-is-a-digital-twin (accessed 3.6.25).

IRS, 2025. Digital Assets | Internal Revenue Service [WWW Document]. URL www.irs.gov/filing/digital-assets (accessed 7.24.25).

Jackson, T., 1999. When Gold Makes Cents: It may Sound Crazy, but the eGold Payment Mechanism Based on Deposits of Precious Metal, is Cheap, Efficient and Easy: [London edition]. *Financ. Times*, 18.

JLT, 2010. Financial Instruments and Exchange Act – English – Japanese Law Translation [WWW Document]. URL www.japaneselawtranslation.go.jp/en/laws/view/2355/en (accessed 7.26.25).

Johnson, B.D., 1984. Discretionary Commodity Accounts as Securities: An Application of the Howey Test. *Fordham Rev.* 53, 639.

Kamel, M.A., Bakhoum, E.S., Marzouk, M.M., 2023. A Framework for Smart Construction Contracts using BIM and Blockchain. *Sci. Rep.* 13, 10217. https://doi.org/10.1038/s41598-023-37353-0

Kloosterman, R., Smits, M., Haji, W., Li, J., 2024. Blockchain-based Common Data Environments to Address Data and Information Fragmentation in The Dutch Construction Industry, in: *Computing in Construction. Presented at the 2024 European Conference on Computing in Construction, European Council for Computing in Construction.* https://doi.org/10.35490/ec3.2024.184

Kouhizadeh, M., Saberi, S., Sarkis, J., 2021. Blockchain Technology and the Sustainable Supply Chain: Theoretically Exploring Adoption Barriers. *Int. J. Prod. Econ.* 231, 107831. https://doi.org/10.1016/j.ijpe.2020.107831

Kouhizadeh, M., Sarkis, J., Zhu, Q., 2019. At the Nexus of Blockchain Technology, the Circular Economy, and Product Deletion. *Appl. Sci.* 9, 1712. https://doi.org/10.3390/app9081712

Kun, H., 2021. *Technological Solutions to Guaranteed Wage Payments of Construction Workers in China.* International Labour Organization, HYPERLINK "https://webapps.ilo.org/static/english/intserv/working-papers/wp048/index.html"https://webapps.ilo.org/static/english/intserv/working-papers/wp048/index.html#:~:text=and%20analyses%20how%20the%20IT,and%20robust%20labour%20inspection%20systems

Latham, M., 1994. Constructing the Team. HM Stationery Office.

LEAP:IN, 2022. Trusted Tech and Things with Twins [WWW Document]. LEAP:IN. URL www.insights.onegiantleap.com/newsletters/trusted-tech-and-things-twins/ (accessed 3.6.25).

Leow, A., 2024. Hype Cycle for Web3 and Blockchain, 2024 [WWW Document]. Gartner. URL www.gartner.com/en/documents/5623191 (accessed 1.18.26).

Library of Congress, 2022. Regulation of Cryptocurrency Around the World, Law Library of Congress (U.S.). Global Legal Research Directorate.

Low, K.F.K., 2020. Bitcoins as Property: Welcome Clarity? https://doi.org/10.2139/ssrn.3538956

Lucking, D., Aravind, V., 2020. Cryptocurrency as a Commodity: The CFTC's Regulatory Framework. https://docslib.org/doc/11514391/cryptocurrency-as-a-commodity-the-cftcs-regulatory-framework (accessed 24.07.25).

Lustenberger, M., 2023. Organizing projects with blockchain through a decentralized autonomous organization. 4, Pages 100102. Project Leadership and Society. DOI 10.1016/j.plas.2023.100102 www.sciencedirect.com/science/article/pii/S2666721523000236 (accessed 09.01.2026).

Maciel, A., 2025. Construction Blockchain [WWW Document]. Construction Blockchain. URL www.constructionblockchain.org (accessed 5.4.25).

Mantalara, G., 2021. An Overview of the ML/TF Risks and Regulatory Responses in the Crypto-asset Landscape. *J. Int. Bank. Law Regul.* 36(11), 487–492.

Marinelli, M., Janardhanan, M.N., 2023. The Value Proposition of Machine Learning in Construction Management: Exploring the Trends in Construction 4.0 and Beyond. pp. 247–272. https://doi.org/10.4018/978-1-6684-5643-9.ch010

Mason, J., Escott, H., 2018. Smart Contracts in Construction: Views and Perceptions of Stakeholders. *Proceedings of FIG Conference*, Istanbul, May 2018.

McKendrick, J., 2024. Digital Twin Efforts Best Shared, US Government Agencies Told. RTInsights. URL www.rtinsights.com/digital-twin-efforts-best-shared-us-government-agencies-told/ (accessed 7.20.25).

McKinsey, 2023. What is Web3 Technology (And Why is it Important)? | McKinsey [WWW Document]. What Web3. URL www.mckinsey.com/featured-insights/mckinsey-explainers/what-is-web3 (accessed 6.1.25).

Mehra, A., 2022. Blockchain Market Worth $67.4 Billion by 2026 – Report by MarketsandMarkets [WWW Document]. GlobeNewswire News Room. URL www.globenewswire.com/news-release/2022/07/04/2473551/0/en/Blockchain-Market-worth-67-4-billion-by-2026-Report-by-MarketsandMarkets.html (accessed 4.15.25).

Morris, N., 2018. Construction Firm Probuild Adopts Brickschain for Blockchain. Ledger Insights – Blockchain Enterprises. URL www.ledgerinsights.com/construction-firm-probuild-brickschain-blockchain/ (accessed 6.1.25).

Mullican, K., 2022. How Digital Twins Will Help Manufacturers Meet Sustainability Goals. Cloud Wars. URL https://cloudwars.com/cloud/how-digital-twins-will-help-manufacturers-meet-sustainability-goals/ (accessed 7.20.25).

Mukhopadhyay, M., 2018. Ethereum Smart Contract Development: Build blockchain-based decentralized applications using solidity. Packt Publishing, Birmingham. HYPERLINK "http://www.oreilly.com/library/view/ethereum-smart-contract/9781788473040/" http://www.oreilly.com/library/view/ethereum-smart-contract/9781788473040/

Nakamoto, S., 2008. Bitcoin: A Peer-to-Peer Electronic Cash System. SSRN 3440802.

Narayanan, A., Bonneau, J., Felten, E., Miller, A., Goldfeder, S., 2016. *Bitcoin and Cryptocurrency Technologies: A Comprehensive Introduction*, Illustrated edition. ed. Princeton University Press.

Nelson, D., 2021. World Economic Forum Shares Roadmap for Deploying Blockchains in Real World [WWW Document]. CoinDesk. URL www.coindesk.com/markets/2020/04/28/world-economic-forum-shares-roadmap-for-deploying-blockchains-in-real-world (accessed 7.26.25).

Oesterreich, T.D., Teuteberg, F., 2016. Understanding the Implications of Digitisation and Automation in the Context of Industry 4.0: A Triangulation Approach and

Elements of a Research Agenda for the Construction Industry. *Comput. Ind.* 83, 121–139. https://doi.org/10.1016/j.compind.2016.09.006
Okta, 2024. Decentralized Identity: The Future of Digital Identity Management [WWW Document]. URL www.okta.com/blog/2021/01/what-is-decentralized-identity/ (accessed 6.1.25).
Omagari, T., Sako, Y., 2019. Japan's New Crypto Regulation: 2019 Amendments to Payment Services Act and Financial Instruments and Exchange Act of Japan. Practice.
PBC, 2021. Notice on Further Preventing and Resolving the Risks of Virtual Currency Trading and Speculation [WWW Document]. URL www.pbc.gov.cn/en/3688253/3689012/4353814/index.html (accessed 7.26.25).
PBC, 2017. Public Notice of the PBC, CAC, MIIT, SAIC, CBRC, CSRC and CIRC on Preventing Risks of Fundraising Through Coin Offering [WWW Document]. URL www.pbc.gov.cn/english/130721/3377816/index.html (accessed 7.26.25).
PBC, 2016. China's Anti-Money Laundering Work: A Decade of Progress [WWW Document]. URL www.pbc.gov.cn/en/3688110/3688175/3707903/index.html (accessed 7.26.25).
Perera, S., Nanayakkara, S., Rodrigo, M.N.N., Senaratne, S., Weinand, R., 2020. Blockchain Technology: Is It Hype or Real in the Construction Industry? *J. Ind. Inf. Integr.* 17, 100125. https://doi.org/10.1016/j.jii.2020.100125
Perez, Y.B., 2019. Gartner Says Blockchain is Sliding into 'The Trough of Disillusionment' [WWW Document]. TNW Fintech-Ecommerce. URL https://thenextweb.com/news/gartner-says-blockchain-is-sliding-into-the-trough-of-disillusionment (accessed 6.1.25).
Pham, H.C., Tran, S.V.-T., Bao, L., 2025. Toward Construction Payment Automation by Integrating BIM and Blockchain Technology. *Transp. Res. Procedia* 85, 143–149. https://doi.org/10.1016/j.trpro.2025.03.144
Poon, J., Dryja, T., 2015. The Bitcoin Lightning Network: Scalable Off-Chain Instant Payments. http://lightning.network/lightning-network-paper-DRAFT-0.5.pdf
Poon, J., Dryja, T., n.d. The Bitcoin Lightning Network: January 14, 2016.
PSA, 2019. Payment Services Act – English – Japanese Law Translation [WWW Document]. URL www.japaneselawtranslation.go.jp/en/laws/view/3078/en (accessed 7.26.25).
Radman, K., Jelodar, M.B., Lovreglio, R., Ghazizadeh, E., Wilkinson, S., 2025. Real-Time tracking and analysis in construction projects: A RealCONs framework. Advanced Engineering Informatics 67, 103511. https://doi.org/10.1016/j.aei.2025.103511
Rao, S., 2021. Mt. Gox – The Fall of a Giant, in: Corbet, S. (Ed.), *Understanding Cryptocurrency Fraud: The Challenges and Headwinds to Regulate Digital Currencies*. De Gruyter, pp. 71–82.
Rashid, A.M.A., 2024. Blockchain Technology in Higher Education: A Comprehensive Review of Applications and Challenges. *Int. J. Res. Publ. Rev.* 5, 1266–1276.
Reiff, N., Chavarria, A., Munichiello, K., 2024. Which Industries Will Blockchain Technology Disrupt Next? [WWW Document]. Investopedia. URL www.investopedia.com/tech/which-industries-will-blockchain-disrupt-next/ (accessed 3.6.25).
Reveley, W., 1791. Panopticon. Wikipedia.
Sadri, H., Yitmen, I., Tagliabue, L.C., Westphal, F., Tezel, A., Taheri, A., Sibenik, G., 2023. Integration of Blockchain and Digital Twins in the Smart Built Environment Adopting Disruptive Technologies—A Systematic Review. *Sustainability* 15, 3713. https://doi.org/10.3390/su15043713

Saxena, N., 2024. Powering the Digital Public Goods with Blockchain Revolution [WWW Document]. URL www.innoraft.com/blog/role-blockchain-digital-public-goods-use-cases-and-innovations (accessed 3.24.25).
Saxena, N., 2024. The Role of Blockchain in Digital Public Goods: Use Cases and Innovations [WWW Document]. InnoRaft. URL www.innoraft.com/blog/role-blockchain-digital-public-goods-use-cases-and-innovations (accessed 3.6.25).
SEC, 2024. SEC.gov | SEC Charges Kim Kardashian for Unlawfully Touting Crypto Security [WWW Document]. URL www.sec.gov/newsroom/press-releases/2022-183 (accessed 7.25.25).
SEC, 1975. Securities Exchange Act – 11935. *SEC Docket* 8, 808.
Sherman, A.T., Javani, F., Zhang, H., Golaszewski, E., 2019. On the Origins and Variations of Blockchain Technologies. *IEEE Secur. Priv.* 17, 72–77. https://doi.org/10.1109/MSEC.2019.2893730
Shishehgarkhaneh, M.B., Moehler, R.C., Moradinia, S.F., 2023. Blockchain in the Construction Industry Between 2016 and 2022: A Review, Bibliometric, and Network Analysis. *Smart Cities* 6, 819–845. https://doi.org/10.3390/smartcities6020040
Shojaei, A., Naderi, H., 2024. Blockchain Technology for a Circular Built Environment, in: De Wolf, C., Çetin, S., Bocken, N.M.P. (Eds.), *A Circular Built Environment in the Digital Age*. Springer International Publishing, pp. 213–228. https://doi.org/10.1007/978-3-031-39675-5_12
Shrouf, F., Ordieres, J., Miragliotta, G., 2014. Smart Factories in Industry 4.0: A Review of the Concept and of Energy Management Approached in Production Based on the Internet of Things Paradigm, in: *2014 IEEE International Conference on Industrial Engineering and Engineering Management. Presented at the 2014 IEEE International Conference on Industrial Engineering and Engineering Management*, pp. 697–701. https://doi.org/10.1109/IEEM.2014.7058728
Sin, D., 2024a. OCBC Partners LTA to Pilot Blockchain-based Conditional Payment Solution for Construction Projects | OCBC Bank [WWW Document]. URL www.ocbc.com/group/media/release/2024/ocbc-partners-lta-to-pilot-blockchain-based-conditional-payment-solution-for-construction-projects (accessed 7.26.25).
Sin, D., 2024b. OCBC Partners LTA to Pilot Blockchain-based Conditional Payment Solution for Construction Projects | OCBC Bank [WWW Document]. URL www.ocbc.com/group/media/release/2024/ocbc-partners-lta-to-pilot-blockchain-based-conditional-payment-solution-for-construction-projects (accessed 6.1.25).
Singh, A.K., Kumar, V.R.P., Irfan, M., Mohandes, S.R., Awan, U., 2023. Revealing the Barriers of Blockchain Technology for Supply Chain Transparency and Sustainability in the Construction Industry: An Application of Pythagorean FAHP Methods. *Sustainability* 15, 10681. https://doi.org/10.3390/su151310681
Skanska, 2019. New Listing: Digital Home Buying Powered by Blockchain [WWW Document]. Skanska – Global Corporate Website. URL https://group.skanska.com/media/articles/new-listing-digital-home-buying-powered-by-blockchain/ (accessed 6.1.25).
Smith, S., 2019. Exploring the Potential of Blockchain Technology for the UK Construction Industry – 2019. WCC, https://constructorscompany.org.uk/wp-content/uploads/2019/05/Exploring-the-potential-of-Blockchain-technology-for-the-UK-Construction-industry-2019.pdf
State Council of the People's Republic of China, 2020. Regulation on Telecommunications of the People's Republic of China [WWW Document]. URL www.lawinfochina.com/display.aspx?lib=law&id=1667 (accessed 7.26.25).

St.Clair, J., 2021. Digital Twins and Self-Sovereign Identity: Build the Next Generation of Simulation with Privacy Preservation. IoT Slam®. URL https://iotslam.com/session/digital-twins-and-self-sovereign-identity-build-the-next-generation-of-simulation-with-privacy-preservation/ (accessed 7.20.25).

Tapscott, D., 2016. Blockchain Revolution: How the Technology Behind Bitcoin Is Changing Money, Business, and the World. Portfolio, New York.

Tapscott, D., Vargas, R.V., 2019. How Blockchain Will Change Construction. Harvard Business Review.

The Economist, 2016. The DAO of Accrue. The Economist.

Tian, Y., Lu, Z., Adriaens, P., Minchin, R.E., Caithness, A., Woo, J., 2020. Finance Infrastructure Through Blockchain-based Tokenization. *Front. Eng. Manag.* 7, 485–499. https://doi.org/10.1007/s42524-020-0140-2

Tian, Y., Minchin, R.E., Petersen, C., Moayed, E., Adriaens, P., 2022. Financing Public-Private Partnership Infrastructure Projects Through Tokenization-enabled Project Finance on Blockchain. *IOP Conf. Ser. Mater. Sci. Eng.* 1218, 012027. https://doi.org/10.1088/1757-899X/1218/1/012027

Tornatzky, L.G., Fleischer, M., Chakrabarti, A.K., 1990. *The Processes of Technological Innovation*, Issues in Organization and Management Series. Lexington Books.

TraceX, 2022. Carbon Traceability on the Blockchain.

Treasury, H.M., 2018. Cryptoassets Taskforce: Final Report. Open Government Licence Bank England, UK London. (accessed 11.19).

Treiblmaier, H., 2018. The Impact of the Blockchain on the Supply Chain: A Theory-Based Research Framework and a Call for Action. *Supply Chain Manag.* 23, 545–559. https://doi.org/10.2139/ssrn.3224145

Tsuchiya, Y., Hiramoto, N., 2021. How Cryptocurrency is Laundered: Case Study of Coincheck Hacking Incident. *Forensic Sci. Int. Rep.* 4, 100241. https://doi.org/10.1016/j.fsir.2021.100241

Turk, Ž., 2020. Interoperability in Construction – Mission Impossible? *Dev. Built Environ.* 4, 100018. https://doi.org/10.1016/j.dibe.2020.100018

Turk, Z., Klinc, R., 2017. Potentials of Blockchain Technology for Construction Management, 196, 638-645. Procedia Engineering. DOI HYPERLINK http://doi.org/10.1016/j.proeng.2017.08.052" 10.1016/j.proeng.2017.08.052 http://linkinghub.elsevier.com/retrieve/pii/S187770581733179X (accessed 24.11.2017).

UK GOV, 2017. UK Digital Strategy 2017 [WWW Document]. GOV.UK. URL www.gov.uk/government/publications/uk-digital-strategy/uk-digital-strategy (accessed 7.24.25).

UNFCCC, 2021a. The Good, The Bad and The Blockchain [WWW Document]. URL https://unfccc.int/news/the-good-the-bad-and-the-blockchain (accessed 3.6.25).

UNFCCC, 2021b. The Good, The Bad and The Blockchain [WWW Document]. URL https://unfccc.int/news/the-good-the-bad-and-the-blockchain (accessed 3.24.25).

U.S. Government, 1936. uscode1958-001007001.

Vadgama, N., Tasca, P., 2021. An Analysis of Blockchain Adoption in Supply Chains Between 2010 and 2020. *Front. Blockchain* 4. https://doi.org/10.3389/fbloc.2021.610476

Valero, E., Bosché, F., Bueno, M., 2022. Laser scanning for BIM. Journal of Information Technology in Construction (ITcon) 27, 486–495. https://doi.org/10.36680/j.itcon.2022.023

Vos, G., 2019. The Launch of the Legal Statement on the Status of Cryptoassets and Smart Contracts. *Judiciary* 25, 2–1.

WEF, 2022. Are DAOs the Business Structures of the Future? [WWW Document]. World Economic Forum. URL www.weforum.org/stories/2022/06/are-dao-the-business-structures-of-the-future/ (accessed 6.1.25).

Werbrouck, J., Schulz, O., Oraskari, J., Mannens, E., Pauwels, P., Beetz, J., 2023. A Generic Framework for Federated CDEs Applied to Issue Management. *Adv. Eng. Inform.* 58, 102136. https://doi.org/10.1016/j.aei.2023.102136

Wood, D.G., 2025. Ethereum: A Secure Decentralised Generalised Transaction Ledger. https://ethereum.github.io/yellowpaper/paper.pdf

Wood, G., 2014. Ethereum: A Secure Decentralised Generalised Transaction Ledger. *Ethereum Proj. Yellow Pap.* 151 , 1–32.

Xiao, J., Zhang, W., Zhong, R.Y., 2023. Blockchain-enabled Cyber-physical System for Construction Site Management: A Pilot Implementation. *Adv. Eng. Inform.* 57, 102102. https://doi.org/10.1016/j.aei.2023.102102

Xiong, X., Luo, J., 2024. Global Trends in Cryptocurrency Regulation: An Overview. https://doi.org/10.48550/arXiv.2404.15895

Xu, Y., Chi, M., Chong, H.-Y., Lee, C.-Y., Chen, K., 2024. When BIM Meets Blockchain: A Mixed-methods Literature Review. *J. Civ. Eng. Manag.* 30, 646–669. https://doi.org/10.3846/jcem.2024.21638

Ye, X., Zeng, N., Tao, X., Han, D., König, M., 2024. Smart Contract Generation and Visualization for Construction Business Process Collaboration and Automation: Upgraded Workflow Engine. American Society of Civil Engineers, 38, Pages 04024030, https://ascelibrary.org/doi/10.1061/JCCEE5.CPENG-5938 (accessed 09.01.26).

Zamyatin, A., Harz, D., Lind, J., Panayiotou, P., Gervais, A., Knottenbelt, W., 2019. XCLAIM: Trustless, Interoperable, Cryptocurrency-Backed Assets, in: *2019 IEEE Symposium on Security and Privacy (SP). Presented at the 2019 IEEE Symposium on Security and Privacy (SP)*, pp. 193–210. https://doi.org/10.1109/SP.2019.00085

Zhang, G., Chen, S.C.-I., Yue, X., 2024. Blockchain Technology in Carbon Trading Markets: Impacts, Benefits, and Challenges—A Case Study of the Shanghai Environment and Energy Exchange. *Energies* 17, 3296. https://doi.org/10.3390/en17133296

Zhang, W., Wu, L., Ji, L., 2025. Blockchain Technology Adoption Strategies for the Shipping Costs Bearer in the Fresh Product Supply Chain. *Front. Sustain. Food Syst.* 9. https://doi.org/10.3389/fsufs.2025.1550985

Zheng, R., Jiang, J., Hao, X., Ren, W., Xiong, F., Ren, Y., 2019. bcBIM: A Blockchain-Based Big Data Model for BIM Modification Audit and Provenance in Mobile Cloud. Mathematical Problems in Engineering 2019, 5349538. https://doi.org/10.1155/2019/5349538

Zheng, Z., Xie, S., Dai, H.N., Chen, X., Wang, H., 2018. Blockchain Challenges and Opportunities: A Survey. *Int. J. Web Grid Serv.* 14, 352. https://dois.org/10.1504/IJWGS.2018.095647

Zhuk, A., 2025. Beyond the Blockchain Hype: Addressing Legal and Regulatory Challenges. *SN Soc. Sci.* 5, 11. https://doi.org/10.1007/s43545-024-01044-y

Ziegler, C., Welpe, I., 2022. A Taxonomy of Decentralized Autonomous Organizations. International Conference on Information Systems.

Zohar, A., 2015. Bitcoin: Under the Hood. *Commun. ACM* 58, 104–113.

# Acronyms

- 5G – See Fifth Generation Mobile Network
- AEC – See Architecture, Engineering, and Construction
- AECO – See Architecture, Engineering, Construction, and Operations
- AI – See Artificial Intelligence
- API – See Application Programming Interface
- BIM – See Building Information Modelling
- BSI – See British Standards Institution
- CBC – See Construction Blockchain Consortium
- CM – See Construction Management
- DAO – See Decentralised Autonomous Organisation
- DAPO – See Decentralised Autonomous Project Organisation
- DeFi – See Decentralised Finance
- DLT – See Distributed Ledger Technology
- ESG – See Environmental, Social, and Governance
- ERP – See Enterprise Resource Planning
- IAM – See Identity and Access Management
- ICT – See Information and Communication Technology
- IoT – See Internet of Things
- KYC – See Know Your Customer
- ML – See Machine Learning
- NFT – See Non-Fungible Token

- PoS – See Proof of Stake
- PoW – See Proof of Work
- TPS – See Transactions Per Second
- ZKP – See Zero-Knowledge Proof

# Glossary of Terms

- **Application Programming Interface** – Enables different software systems to communicate.
- **Architecture, Engineering, and Construction** – Disciplines in the built environment industry.
- **Architecture, Engineering, Construction, and Operations** – The broader built environment sector encompassing the full lifecycle of the built environment.
- **Artificial Intelligence** – Systems that simulate human intelligence; in construction, used for predictive analytics, automation, and decision-making.
- **Audit Trail** – Immutable, chronological record of transactions or events.
- **Blockchain** – A decentralised, tamper-evident ledger recording transactions across a network.
- **British Standards Institution** – The UK's national standards body, responsible for publishing technical standards.
- **Building Information Modelling** – A digital representation of a facility's physical and functional characteristics that enhances planning and project delivery.
- **Carbon Passport** – A verified digital record of a product's carbon content and life-cycle emissions.
- **Construction Blockchain Consortium** – An organisation promoting blockchain innovation in the built environment sector.
- **Construction Management** – A professional discipline (and occasionally a journal) focused on planning, coordination, and control of a construction project from inception to completion.

- **Construction 4.0** – The digital transformation of the construction industry through technologies such as AI, IoT, robotics, and blockchain.
- **Consortium Blockchain** – A blockchain governed jointly by a group of known organisations, typically used for enterprise applications.
- **Compliance-by-Design** – An architectural principle embedding legal and regulatory requirements into digital system and workflow design.
- **Cryptographic Techniques** – Methods used to secure communication and ensure the integrity and confidentiality of digital data.
- **Data Ontology** – Structured categorisation of data elements to support consistent interpretation and integration.
- **Data Ownership** – Legal and practical rights associated with the control, access, and use of data in blockchain systems.
- **Data Sovereignty** – The principle that data is subject to the laws and governance structures of the jurisdiction in which it is collected or stored.
- **Data Standardisation** – The process of applying common formats, structures, and protocols to data to enable interoperability.
- **Decentralised Autonomous Organisation** – A blockchain-based organisation managed by encoded rules in smart contracts, without central leadership.
- **Decentralised Autonomous Project Organisation** – A DAO variant specifically structured to manage and govern construction projects.
- **Decentralised Finance** – A blockchain-based financial system that operates without central financial intermediaries, relying instead on smart contracts.
- **Digital Identity / Self-Sovereign Identity (SSI)** – A model allowing individuals or organisations to control their digital identities independently, enhancing privacy and user consent.
- **Digital Resilience** – The ability of a system to recover from, adapt to, or withstand digital threats or disruptions.
- **Digital Signatures** – Cryptographic methods that verify the authenticity and integrity of digital content, documents, or transactions.
- **Digital Twin** – A dynamic, digital representation of a physical asset or process that reflects real-time operational data.
- **Digital Wallet** – A secure software application used to store, manage, and sign blockchain-based digital assets.

- **Dispute Resolution Mechanisms** – Defined processes (on- or off-chain) to resolve disagreements related to digital transactions, smart contracts, or governance.
- **Distributed Ledger Technology** – A digital database that is consensually shared and synchronised across multiple sites and institutions without a central administrator.
- **Electronic Signature / Digital Signature** – A legally recognised cryptographic means of confirming identity and content integrity in digital transactions.
- **Enterprise Resource Planning** – Software systems for managing and integrating core business processes, often linked to blockchain and BIM in construction.
- **Enterprise System Integration** – The linking of blockchain platforms with other enterprise technologies such as ERP, BIM, and IoT.
- **Environmental, Social, and Governance** – A sustainability and ethics framework guiding organisational behaviour, increasingly supported by verifiable blockchain audit trails.
- **ESG Reporting** – Disclosure of non-financial performance indicators such as carbon emissions, social impact, and governance practices.
- **Ethereum** – A decentralised blockchain platform that introduced the concept of programmable smart contracts.
- **Executive Sponsorship** – Strategic support and leadership commitment to technology adoption and digital transformation.
- **Fifth Generation Mobile Network** – Advanced cellular technology providing high bandwidth, low latency, and support for dense IoT deployments.
- **GDPR (General Data Protection Regulation)** – European Union regulation concerning data protection and privacy, with implications for blockchain's immutability.
- **Governance** – The structures and rules that define decision-making authority, accountability, and rights within systems or organisations.
- **Identity and Access Management** – Systems and policies for controlling access to data and resources based on identity verification.
- **Information and Communication Technology** – The suite of digital tools and infrastructure enabling information processing and communication.

- **Intellectual Property** – Legal protections for creative and intellectual works, including digital assets and parametric models on blockchain.
- **Interoperability** – The ability of different systems, platforms, or organisations to work together through compatible data exchange and processes.
- **Internet of Things** – A network of physical devices embedded with sensors and connectivity, enabling them to collect and exchange data.
- **Jurisdictional Risk** – The legal uncertainty or exposure faced by systems operating across different national or regional regulatory environments.
- **Know Your Customer** – A regulatory requirement involving identity verification in financial and compliance processes, also applicable to blockchain applications.
- **Layer 2** – Secondary blockchain protocols designed to enhance transaction speed and scalability without compromising the base layer's security.
- **Liability Allocation** – The contractual distribution of legal responsibilities among parties, especially in decentralised and automated systems.
- **Machine Learning** – A subset of AI that enables systems to learn patterns from data and improve over time without being explicitly programmed.
- **Material Passport** – A digital record of a building material's properties and lifecycle, supporting reuse and circular economy practices.
- **Middleware** – Software that connects different systems or applications, facilitating interoperability between blockchain and legacy systems.
- **MiCA (Markets in Crypto-Assets Regulation)** – An EU regulatory framework aimed at creating a unified approach to the oversight of crypto-assets and service providers.
- **Non-Fungible Token** – A unique blockchain-based digital asset that is not interchangeable with others, used for art, assets, and identity.
- **Oracle** – A service or protocol that delivers real-world data to smart contracts on a blockchain, enabling trustless external input.
- **Payment Automation** – The use of smart contracts to trigger, execute, and record payments without manual intervention.
- **Permissioned Access / Role-Based Access Control (RBAC)** – Systems that restrict data or process access based on predefined user roles and permissions.

- **Pre-Bitcoin Experiments** – Early digital currency initiatives like DigiCash and e-Gold that laid the groundwork for cryptocurrencies.
- **Private Blockchain** – A permissioned blockchain where access is restricted to selected participants, typically used in enterprise contexts.
- **Privacy-Preserving Computation** – Cryptographic methods such as zero-knowledge proofs or SMPC that protect user data while maintaining transparency and auditability.
- **Project Management** – The planning, coordination, and oversight of tasks and resources to meet defined objectives, often enhanced through digital tools.
- **Proof of Stake** – A consensus mechanism where validators are selected based on the amount of cryptocurrency they hold and are willing to "stake".
- **Proof of Work** – A consensus mechanism where participants solve complex mathematical problems to validate transactions and secure the network.
- **Public Blockchain** – A fully decentralised blockchain where anyone can participate and validate transactions.
- **Public-Key Cryptography** – A method that uses paired public and private keys for secure communication and identity verification.
- **Reality Capture and Reality Mesh** – Technologies like laser scanning and photogrammetry that convert real-world environments into digital 3D models.
- **Regulatory Compliance** – The act of aligning systems and processes with applicable laws, standards, and regulations.
- **Regulatory Sandboxes** – Controlled environments established by regulators to allow safe experimentation with new technologies, including blockchain.
- **Scalability** – The capability of a blockchain network to handle increased demand, users, and transactions efficiently.
- **Smart Contract** – A self-executing agreement encoded on a blockchain, with execution triggered by predefined conditions.
- **Smart Contract Legal Validity** – The assessment of whether smart contracts meet legal criteria for enforceability such as mutual consent and consideration.

- **Stablecoin** – A cryptocurrency designed to maintain stable value by being pegged to fiat currency or other assets.
- **Stakeholder Alignment** – The process of ensuring that project participants share goals, expectations, and responsibilities.
- **Supply Chain Transparency** – The ability to trace and verify the origin, status, and movement of materials and products through blockchain.
- **Token Classification** – Legal determination of a digital token's nature (e.g., utility, security, payment), affecting how it is regulated.
- **Token Model** – A framework for the creation and use of blockchain tokens to represent assets, rights, or access.
- **Tokenisation** – The process of converting physical or digital assets into blockchain-based tokens.
- **Training and Upskilling** – Organisational efforts to equip personnel with the competencies required to use and manage new technologies.
- **Transactions Per Second** – A metric indicating how many operations a blockchain network can process within a second.
- **Trustless Interactions** – System-mediated transactions where no party needs to trust another or an intermediary.
- **Unalterable Transactions** – Records stored on blockchain that cannot be changed once confirmed, ensuring auditability.
- **Unlawful Data Immutability** – Legal tension arising when blockchain immutability contradicts rights to delete or alter data under regulations like GDPR.
- **Zero-Knowledge Proof** – A cryptographic technique enabling one party to prove knowledge of information without revealing the information itself.

# Further Reading and Resources

Kassem, M., Maciel, A., Hall, D.M. (Eds.), 2025. *Blockchain, Smart Contracts and Distributed Ledger Technologies in the Built Environment: Key Concepts, Technologies, and Applications*, 1st ed. The Institution of Engineering and Technology, S.l.

Sawhney, A., Riley, M., Irizarry, J. (Eds.), 2020. *Construction 4.0: An Innovation Platform for the Built Environment*, 1st ed. Routledge.

# Index